TENACIOUS BEASTS

TENACIOUS BEASTS

Wildlife Recoveries That Change How We Think about Animals

CHRISTOPHER J. PRESTON

The MIT Press
Cambridge, Massachusetts
London, England

The MIT Press would like to thank the anonymous peer reviewers who provided comments on drafts of this book. The generous work of academic experts is essential for establishing the authority and quality of our publications. We acknowledge with gratitude the contributions of these otherwise uncredited readers.

This book was set in Capitolium2 and URW DIN Condensed by the MIT Press. Printed and bound in the United States of America.

Library of Congress Cataloging-in-Publication Data

Names: Preston, Christopher J. (Christopher James), 1968- author.
Title: Tenacious beasts : wildlife recoveries that change how we think about animals / Christopher J. Preston.
Description: Cambridge, Massachusetts : The MIT Press, [2023] | Includes bibliographical references and index. | Summary: "Conventional wisdom is that wild animals are being wiped out. But conventional wisdom skips some important details. Wildlife is rebounding. Not everywhere. Not every species. But a handful of wildlife populations have reached numbers unimaginable in a century. Red deer in Europe, bison in North America, humpback whales in the Atlantic. They have all seen their populations explode. They are back from the brink, numbering in the tens, or even hundreds, of thousands. Their return thrills those who have rooted for their recovery. It terrifies those who grew comfortable without them. This book tracks—and tries to understand—these dramatic rebounds. It shines a light on species returning to forests and farms, prairies and oceans, rivers and cities. It asks how these transformations can be happening and what they have to teach"—Provided by publisher.
Identifiers: LCCN 2022011919 (print) | LCCN 2022011920 (ebook) | ISBN 9780262047562 (hardcover) | ISBN 9780262372541 (epub) | ISBN 9780262372558 (pdf)
Subjects: LCSH: Wildlife reintroduction. | Animal populations—Effect of climatic changes on.
Classification: LCC QL83.4 .P74 2023 (print) | LCC QL83.4 (ebook) | DDC 639.97—dc23/eng/20220524
LC record available at https://lccn.loc.gov/2022011919
LC ebook record available at https://lccn.loc.gov/2022011920

10 9 8 7 6 5 4 3 2 1

For my father, Robin Preston,
who taught me to love animals

CONTENTS

RESURGENCE

Standing on the dock in the dead of arctic winter, Svein Anders closed his eyes to the northern lights above. All the professor wanted to do right now was listen. As the mountains and green aurora faded from his retina, the sound of a gently rippling ocean washed over his body. Seconds after his mind quieted, he heard the splash of a large object followed by a fierce exhalation of air. Then he heard another. Then another. A fishy scent wafted shoreward. Casting his head back to the sky, Svein Anders's mouth cracked open as he drank in the frigid, nighttime air. He was perched on the doorstep of whales. Dozens and dozens of whales. Humpback and killer whales drawn to Norway's Kaldfjord in the winter months to feast on herring.

The professor lived on the fjord for its cold beauty. Mountains rose in a wall before his kitchen window. Birch trees formed a stubble around his house. Lingonberry flowers speckled the hills in June. The professor built his home with reused materials, spending hours straightening old nails in order not to buy new. He kept chickens and grew potatoes, raised and slaughtered his own sheep. For income, he taught environmental classes at a nearby university and published opinion pieces in the local paper on wind farms, reindeer herding, and the politics of nearby fisheries. His life ebbed and flowed with the seasons.

But the experience on the fjord that night was new. Whales were back in their hundreds, and his—and his neighbors'—world was being transformed. The professor felt only elation. But the professor was not the only person having a new experience with wildlife that year.

It was well past midnight when Sue Davies woke to the sound of a sharp hiss beside her bed in an English seaside town. She bolted upright and fumbled for the light switch. An animal ran quickly out the bedroom door. Squinting in the brightness, Davies saw the bedcovers pulled to one side and felt a pain in her leg. She looked down and saw a small patch of skin starting to swell. She put on bedroom slippers and hobbled downstairs. Her shoes were spread across the kitchen floor and the cat flap lay shattered in the mudroom. Davies had been bitten by a fox.

The first thing Davies thought about was rabies. Bats, feral cats, and foxes all carry the deadly disease. She drove hurriedly to Eastbourne General District Hospital to get the bite mark examined, worrying the emergency room staff would think she was mad. A fox in an upstairs bedroom at 2 a.m. sounds absurd. The doctor inspected her leg and carefully cleaned the wound before sending her home with a tetanus shot. As far as rabies was concerned, Davies needn't have worried. The English Channel keeps infected animals out of the United Kingdom.

Davies went home healthy but alarmed. A wild animal wandering through a modern, centrally heated house is a form of heresy. It challenges the barrier between the safe space of home and the wild world beyond. In the UK, where the landscape is gentle and most of the large animals you see are domesticated, a threat to this barrier is almost a full-scale cultural assault.

Davies's experience turns out to be far from unique. A fox bit twin girls Lola and Isabella Koupparis in their upstairs bedroom in Hackney, Central London. Another entered a home in southeast London and grabbed a four-week-old baby off his bed, nearly severing his finger. A third pushed through a kitchen door in Plymouth and bit a seven-month-old sitting in her bouncer in the kitchen.

Nobody knows for sure why foxes are entering so many homes. But one thing nobody disagrees about is this: in the last couple of decades, people are encountering many more foxes in their daily lives. Their pointed faces are everywhere, poking their noses into rubbish bins, darting through hedgerows, and walking jauntily along the tops of garden walls.

Emboldened foxes and fjords full of whales tell us something. Amid a crushing wave of bad news, a handful of wildlife species have bounced back. The recoveries create glimmers of hope and bring complicated challenges. What they show about animals and their tenacity is inspiring. What they demand from minds unaccustomed to their presence is where some of the deepest puzzles lie. We need, in short, a new way to think about animals.

* * *

The news about animals is bleak. Wildlife populations have declined 20 percent over the last century. More than nine hundred species have been wiped off the face of the earth since industrialization. At least a million are threatened with extinction, many of which now number less than a thousand. Vertebrates are particularly hard hit, with populations plummeting over 60 percent since 1970. Birds in North America are down 29 percent. More than 40 percent of amphibians and a third of ocean corals are in danger.

The collapse is almost certain to get worse. Nearly three-quarters of ice-free land and two-thirds of the oceans have been transformed by human activities. Seventy percent of the world's fresh water is devoted to crop and livestock production. Soil is eroding one hundred times faster than it is being built. The signs of human dominance are everywhere. Ninety-six percent of the weight of the world's mammalian life is either human or domestic animals. Let that sink in. People and their livestock weigh twenty-four times all the elephants, whales, and tigers combined.

Climate change is adding to the problems. Temperatures on land have risen 1.59 degrees Celsius above preindustrial levels, climbing even more at high latitudes. Fifty percent of Australia's Great Barrier Reef dies back each summer due to coral bleaching, a number that's only going to increase. Barely 7 percent of ocean fish stocks are harvested at sustainable rates, and marine waters are becoming warmer and more acidic. The Intergovernmental Panel on Climate Change predicts that only a few years are left before many low-lying island nations, together with the biodiversity they support, will sink beneath the waves.

And yet, despite the grim outlook for wildlife and despite the relentless stream of bad news, some animal populations are on the rebound. Humpback whales in the western Indian Ocean have surged from six hundred in the middle of the last century to over thirty thousand today. Black bears in California have quadrupled in a few decades. Storks are breeding in southern England for the first time since the Middle Ages. Wolves now number more than twelve thousand in Europe, and their range has expanded into densely populated countries, including the Netherlands and Belgium, where they have been absent for a century.

None of this should be happening. With such a dire outlook for wildlife, how is any of it possible?

The pages ahead probe the mystery by looking at a number of wildlife recoveries. Each section tracks one or more animals and reveals how their numbers have bucked the larger trend. It asks how things can go so right when so much is going so wrong.

The focus is mainly on species in the United States and Europe because these are the places I know best. One is the continent on which I grew up. The other is the continent on which I am growing old. I don't claim anything I say about animal recoveries is unique to these locations. I'm confident the lessons are available elsewhere. There are probably clearer or more illuminating cases to explore. But I focus on where I have lived and worked, for these are the places most familiar to me and where my affections are most fierce.

The recoveries are not hard to spot—roe deer rutting in the forest where I grew up in southern England, pronghorn antelope trotting west of the Continental Divide near my home in Montana, seabound trout flicking their tails past dismantled dams in Washington state, bears crunching acorns in Italy's Apennines. When you scratch the surface and turn a few stones, you find animals everywhere, waiting for a chance to return.

Highlighting the recoveries is not to question the established wisdom about species loss. That truth cannot be doubted. For every success, there are ten cases of continuing decline. Enormous challenges lie ahead in the form of climate warming, increasing consumption, the spread of non-native species, and habitat loss. Some of the recoveries I discuss are tentative. A few may never get off the ground at all. So let me be absolutely clear from the start because I want no possibility of misunderstanding. This book is not designed to provide soothing reassurances about wildlife. Animal populations are not out of danger. Their outlook remains dire.

But there are reasons to examine the fragments of good news. The first is to provide hope. Amid growing mountains of loss, some species have shown the tenacity to bounce back. Their recoveries can inspire us. If we create the right conditions, if we offer them a slender chance, we may be surprised by the rapid resurgence of animal life. This is what wildlife does. It seizes its opportunities. Creative and adaptable creatures stand waiting in the wings, hardwired to seize their moment.

Alongside hope, there is another reason to write this story, more philosophical than ecological. The return of animals demands something from us. Wildlife challenge entrenched ways of thinking and shake us from well-worn habits of mind. They teach us better ways of being in the world. This is no longer the nineteenth century, when industrializing nations felt entitled to exploit wildlife as they saw fit. Humans possess a range of new tools and an assortment of new desires. If wild animals are to flourish on twenty-first-century landscapes, a different set of attitudes must take hold. We need a better way to think about wildlife. The chapters ahead contain plenty of ecological details. But they also hint at a philosophy, a roadmap to a future state of mind.

Each section of the book highlights one essential change in attitude that recovering species demand. Often this means a change in how to act. But it can also mean a change in how to think. You should read this book looking for the transformation in attitude recommended in each section. Farmland Villains probes how recovering wolves in Europe challenge the gulf between the wild and the civilized. Prairie Puritans uses the return of bison to ask whether some returning wildlife are more authentic than others and whether this has anything to do with genes. River Engineers discusses salmon, dams, and beavers to pose questions about human expertise with managing complex systems. Forest Managers considers when owls, bears, and wildcats should—and should not—be left alone. Ocean Partners looks at humpback whales and sea otters and asks if they can be thought of as allies in a generational battle. On each landscape, I make the case for a better set of ideas around wildlife. I tease out more suitable ways of thinking about the species that surround us.

Some cultures failed spectacularly to live alongside animals first time around. They wiped them out thoughtlessly and without compassion. I use a handful of wildlife recoveries to ask how things might be different this time—how we might consider animals in ways that let them live. Embracing these fresh states of mind won't be easy. Changing how we think is often harder than changing what we do. But the cost of doing nothing will be the end for many species. And we will encounter some excellent guides along the way.

* * *

On a rural road in Canada a few Octobers ago, I pulled the car over to stare at two long-legged animals loping confidently through an adjacent field. They trotted nonchalantly along, not far from several dozen unconcerned cows. I had never seen a wolf outside a national park before, but I knew instantly what they were. When I lowered my window, the pair of wolves stopped and stared directly at my vehicle. There was an assertiveness in their look. They belonged on the land. As I admired the luster of their coats and the muscled curve of their haunches, I felt my place in the world shift just a notch. The feeling was uplifting but, at the same time, mildly disorienting. I had not expected them there. Something in my mind had to shift. The wolves finally grew tired of looking, bumped shoulders, and resumed their lope across the field. I drove the car back onto the road with electricity coursing through my blood. The feeling did not fade for many miles. It persists in the pages of this book.

FARMLAND VILLAINS

1 CARS AND CARCASSES

Hugh Jansman's work puts him in closer contact with wild, predatory carnivores than any Dutchman in 140 years—closer, at least, to the dead ones. Jansman is a wildlife ecologist at Wageningen University, part of a team responsible for investigating suspicious deaths of animals in the woods, heaths, and farmlands of the Netherlands. There has recently been an uptick in demand for his skills.

The squat, curly-haired Jansman is trained in medicine and anatomy. He is an expert at cutting up dead bodies. He pairs his knowledge of animal physiology with the latest tools in DNA analysis to unearth the truth behind an animal's last moments. It sounds like a gory job spent in close proximity to blood, saliva, and stomach contents. Most of the time, it probably is. But the forty-two-year-old ecologist grins like a five-year-old when you ask him to talk about the new animal that is increasingly on the sharp end of his dissection knife.

Jansman lives in the Veluwe, a large wooded area in the center of the Netherlands. At just over a thousand square kilometers, the mixed landscape feels as much like wilderness as anything found in Holland. The hills and ridges of the Veluwe rest high and dry, pushed upward by advancing Pleistocene glaciers. Like most landscapes in the Netherlands, this does not make them very high or particularly dry. Their mightiest folds reach only 350 feet above the gray waters of the North Sea. The forests are greened by more than thirty inches of rain per year. Even though they lack the soaring beauty of the Dolomites or the Pyrenees, Jansman loves it here. He loves it even more now that this wild heart of the Netherlands has gained a remarkable new resident.

In the summer of 2019, two wolves paired up in the Veluwe and gave birth to a litter of pups. The Netherlands had its first resident wolf pack in a century and a half. Among well-ordered tulip fields, picturesque windmills, and grass-covered dikes, the arrival of a wolf pack was rattling. Wolves padding through the Veluwe didn't just signal a notable ecological change. It prompted an urgent confrontation about what it means to be Dutch.

* * *

Wolves vanished from the Netherlands in the mid-nineteenth century. The truth is, they never stood a chance. Farmers saw them as a threat to their livestock. Hunters blamed them for declines in deer. They were massacred by the hundreds with traps, poison, and guns. The surviving wolves struggled to find prey as the human population grew and forests made way for agriculture.

A carefully crafted anti-wolf mythology stoked the desire for their slaughter. Wolves were evil, sharp-fanged beasts that preyed on children. They were, according to one early commentator, "the very incarnation of destruction, with his powerful jaws of shark teeth . . . and the cunning of man." Villagers feared and then demonized them. Famines were known simply as "the wolf." A painful and disfiguring skin condition was called *lupus vulgaris*. The creature's presence was deemed incompatible with civilized life. Hunters, farmers, and those who held a vendetta went to war on the retreating packs. By 1869, the last Dutch wolf was dead.

Europeans celebrated their demise. The slaughter fit the worldview they occupied. The rationalist philosophy of René Descartes created a perfect space for wolf hatred. Descartes, who published most of his major works while living in the Netherlands, insisted there was a giant metaphysical gulf between humanity and the rest of creation. Humans alone had a mental element to their existence. *Cogito, ergo sum*: I think, therefore I am. Animals were simply machines, stitched together out of flesh and sinew. They had no mind, no experience, no feelings. Descartes thought nothing of experimenting on dogs and cats without anesthesia. A wriggle or a yelp was like a squeaky brake on a car. Everything could be explained mechanically. There was no pain in the animal kingdom. Its residents had no value other than the ways they satisfied human whim.

The chasm Descartes proposed between humans and animals fit neatly alongside the prevailing Christian worldview. Humans sat on a divinely sanctioned pedestal. The rest of creation was there to serve them. The metaphysical gulf suited the scientists and industrialists of the emerging Dutch empire. The natural world was there for the taking, whatever the cost in animal—and often human—blood. Wildlife were persecuted without guilt until well after most of them were gone.

For more than a century and a half, nobody seemed to care about the wolf's absence. Farmers told enough fables about the big, bad wolf that it seemed like a good riddance. Fairy tales collected by the Brothers Grimm were populated with wolves who tricked and hurt humans. Stories like "Little Red Riding Hood" ensured European children grew up with an exaggerated fear of the wolf. The animal itself now existed in memory only. After a couple of generations, the thought of a wolf lurking in the shadows slipped away. Wolves were a haunt from a previous era, a time when windmills were the main source of power and a formidable Dutch navy protected trade routes all the way to Southeast Asia.

* * *

All this sat comfortably with the Dutch until, after more than a century's absence, a wolf carcass showed up unexpectedly on a busy roadside. An early-morning driver found the body of a 110-pound female on the edge of a field near Luttelgeest in Flevoland. The dead wolf appeared to have been hit by a car. Its bone structure was damaged, and flecks of blood on its fur pointed to a high-speed impact of some sort.

The public was flabbergasted. Some thought it was a hoax, similar to the case of a dead seal that had shown up in a field sixty miles from the sea a few years previously. It was hard to imagine a wolf running through the middle of one of the most intensively managed landscapes on earth. Wildlife biologists were elated but a little suspicious. They knew wolf numbers had been climbing across the border in Germany. They understood that Dutch habitat wasn't much different from the habitat to the east. Reports from German farmers and grainy nighttime footage from camera traps confirmed wolves were heading in their direction. Wolf supporters had been holding their breath for Europe's most storied canid to lope across the border into the Netherlands. They had already founded an organization, Wolven in Nederlands, to look after them when they arrived. Officials in the city of Steenwijk collected the body and put it in a cooler while they pondered what to do.

Jansman chuckles from behind his lion's mane of hair when he recalls how there were doubts about the Luttelgeest wolf from the start. It all seemed a bit too sudden. The place where the carcass showed up was a bit odd for the first Dutch wolf in 150 years. It was close to a densely populated district and over sixty miles from the German border. Until then, nobody had reliably confirmed a wolf on the Dutch side of the line. There had been rumors. An engineer on a train from Amsterdam to Almere glimpsed a wolf-like animal trotting away from the tracks. A farmer swore he had seen a wolf crossing his fields, and a walker spotted muddy prints and suspicious-looking scat. But there was no solid evidence. If this new wolf had made it all the way to Flevoland before getting hit, she had managed to keep a remarkably low profile on the way.

The day after the carcass was found, the wolf became Jansman's responsibility. His lab at Wageningen University was instructed to conduct a necropsy with assistance from the Dutch Wildlife Health Center in Utrecht. He drove over to the city of Steenwijk and picked up the carcass. He remembers the surreal feeling of driving the Netherlands' first wolf in over a century across town in the back of his car. He wanted to yell out the window about what he had in the trunk.

Back at the lab, Jansman and his colleagues quickly confirmed it was a wolf and sent out some DNA for analysis. The X-ray machine needed for a full necropsy was broken. But people wanted answers fast. When

Jansman's team started cutting into the wolf, right away they encountered puzzles. The stomach contents showed the last thing the wolf ate was a beaver. When the beaver's partially digested tail was tested, carbon isotopes matched isotopes from eastern Europe and the Baltics, over seven hundred miles away. This was a long way for a wolf to travel between meals.

Jansman and his colleagues expected their investigation to reveal injuries from a car's impact. But when the forensic team looked closely, they discovered damage to the skeleton consistent with a high-velocity projectile. When they finally got hold of an X-ray machine, it revealed metal fragments and signs of a bullet grazing the wolf's spine. They also found matching holes in the skin. The wolf hadn't been killed by a car at all. Someone had shot it, probably in Poland, before driving it across two countries and placing it alongside a Dutch road as some sort of joke.

Those who earlier appeared on TV celebrating the return of the wolf suddenly found themselves looking foolish. Jansman says all the hype had been distracting. A little more patience would have spared some blushes. But the false start served an important purpose. The hoax woke up the Dutch to the fact *Canis lupus* was about to reclaim its place in their country. Young wolves are constantly on the lookout for new territory, and habitat in Germany was filling up fast. Only three months before the carcass appeared in Luttelgeest, a German wildlife camera snapped pictures of a wolf less than ten kilometers from the border—a casual morning stroll for a wolf. Jansman knew it was only a matter of time before a wolf ignored the lines drawn by Europeans on their maps and reclaimed another part of its range.

The Dutch government quickly issued a statement making it clear all arriving wolves would be protected. The European Union's 1992 Habitats Directive safeguarded wolves across their entire natural habitat. This meant EU member states must provide protection anywhere the wolf used to howl. Not only were wolves coming back to the Netherlands. When they did, they would be here to stay.

Right on cue, credible sightings of wolves started to trickle in. Three different eyewitnesses spotted a wolf walking close to the Dutch town of Twente before it apparently returned to Germany. The following year, several people filmed a wolf trotting down the pavement of a housing estate near Groningen. Others recorded footage of the same wolf walking near wind turbines on an industrial estate. The wolf seemed unusually comfortable around people, prompting a headline in the British *Daily Mirror*: "Terrifying footage of wolf prowling city streets looking for its next meal." Jansman, who was carefully monitoring the wolf sightings and matching the photographs with a German database, doubted the *Mirror*'s take. He had his own theory about the unusual behavior. The wolf was likely habituated to humans by German soldiers feeding it on military bases over the border.

Despite the *Daily Mirror*'s alarm, many Dutch people found images of a wolf walking down a street in broad daylight more perplexing than terrifying. The Netherlands doesn't look like a place that can support a wolf pack. The landscape is highly domesticated. Tractors plow straight lines across fields ringed by well-maintained drainage ditches. Flocks of free-range chickens spill out of sheds into neatly fenced rectangular plots. Large blue locomotives marked by the "NS" of the national railway carrier crisscross the countryside on electrified tracks. Bicyclists dart around like honeybees on a trail system that is a source of national pride. Having one of the most iconic symbols of the wild trotting through the middle of all this was bizarre.

If many Dutch people were happy, Jansman was ecstatic. As an ecologist and wildlife enthusiast, he knew the benefits. Dutch hunters and government agencies have to kill nearly 90 percent of the wild boar and two-thirds of the deer in the Veluwe each year to prevent crop damage. The wolves were going to help out wildlife managers. They would kill some deer and teach the rest to be more skittish. Wolves were also going to make the landscape more interesting to people like Jansman. Many Europeans crave the experience of wildlife available in North America, Africa, or Southeast Asia. It was hard to find anything like it at home. There was a reason Jansman chose to live in the Veluwe. Increasing the diversity of local wildlife would add spice to Dutch life.

But Jansman's excitement was also tempered with a strong dose of realism. He knew wolves would complicate things. Learning to live with wolves was returning to a perilous past. Centuries of prejudice stacked the deck against the wolf's return. For most people, wolves still existed on a different plane of being. They were powerful predators with sharp teeth and claws. They slunk around under cover of darkness, stealing what they needed. They killed the animals humans raised for themselves, leaving the carnage undiscovered until dawn.

The challenge presented by the wolf in the Netherlands is a potent illustration of what many animal recoveries demand. To recover a wild animal, you don't just need the right habitat, enough shelter, and sufficient prey. Nor is recovery simply a matter of better practices to allow humans and wildlife to coexist. You need to do more. You need to think differently about what the word *wildlife* really means.

* * *

A nonexpert might wonder initially how wolves have recovered so well. Females are sexually receptive for only one or two weeks each year. A typical litter ranges between four and seven pups, but only half make it into adulthood. The mother's life is devoted almost entirely to the survival of her offspring, who have a lot to learn before they can thrive. For someone who doesn't know wolves, the odds look long. But *Canis lupus* has a trick or two up its sleeve. Wolves are smart and adaptable. They are generalist feeders. Their brains are packed into the body of one of nature's supreme athletes. A complex social structure makes the pack into a well-tuned survival machine, year-in and year-out. Wolves, it turns out, are capable of astonishing comebacks.

Wolves mate in winter or early spring. The pack's breeding pair digs out a den from a hillside or under a tree root. The mother stays active in pack life until she feels the arrival of her young is imminent, at which point she retreats to the den. After sixty-two days of gestation, the pups are born blind and deaf. They nurse every few hours for several weeks. The mother remains in the den with the young and relies on the other adults to bring her food. The hungry pups grow fast, gaining about three pounds a week.

The pups' eyes open after two weeks, always blue at first. At three weeks, their flopped-over ears perk up, and they begin to hear sound. At five weeks, they are strong enough to make their first forays outside. With oversize feet, large heads, and limited coordination, they romp around the mouth of the den play-biting their siblings and tumbling over each other. The littermates, like human siblings, compete playfully to establish dominance, in the process establishing lasting attachments to each other. Both the parents and the pups are open with their affection, cute as any litter of domestic puppies.

At the end of six weeks of nursing, the young wolves are ready to wean. The mother, now comfortable leaving the den for short periods, brings back food to her offspring. The aunts and uncles help out and babysit, a system animal behaviorist Carl Safina calls *extended childcare*. Wolves feed their young by regurgitating meat they have eaten. The food is chewed and partially digested, making the transition from milk easier. This task is usually performed by one of the parents, but other members of the pack participate too. The pups jab and lick at the adults' jaws to stimulate regurgitation. Between eight and sixteen weeks, the pack moves from the den to what can end up being a sequence of rendezvous sites. There they hang out, sleep, and play with toys like sticks, bones, and bits of fur. Through the summer and early fall, the young wolves' high-pitched howls deepen and gain strength. Their legs lengthen, and their frames fill out. The soft blue eyes turn a piercing yellow. At six or seven months, the young are almost indistinguishable from the adults, and they start tagging along on hunting expeditions.

The social structure of the pack helps its members survive the challenges of the months ahead. Hunting is difficult and often dangerous. Research cited by Barry Lopez in his classic study *Of Wolves and Men* found 50 percent of wolves on the Tanana River in Alaska showed signs of traumatic injury from hunting. But success with big prey is rewarding. A wolf's digestive system has evolved for feast or famine. Wolves can eat a fifth of their bodyweight at one sitting.

Like strong human societies, wolves watch out for each other. They hunt and play together, look after the sick, and share food. The breeding pair have a strong bond and lead with a learned and sophisticated pack intelligence. The term *alpha wolves* is no longer in favor. Pack structure is determined less by belligerence than by knowledge and experience. Females are highly respected. Successful hunts are orchestrated by the older wolves, even when they may no longer bring the prey down. The pack values its elders and protects them. Dark lines around the eyes, ears, and muzzle help wolves make expressive gestures of joy, anxiety, or aggression. While wolves for the most part keep aggression in check within the pack, they can be ruthless when confronting outside threats. The greatest source of wolf mortality in Yellowstone National Park is other wolves.

A healthy pack breeds every year. Normally, the pack has only one litter, though when conditions are good, a second or even third female may also breed. The wolves that make it to adulthood reach sexual maturity after two years. A surge of hormones is a signal to find their own territory. Some are killed when they enter the home range of established packs. But wolves are smart, and they avoid trouble when they can. If there is suitable unoccupied land with an adequate prey base, a wolf will find it, sometimes after a journey of hundreds of miles. The constant pressure for adolescents to find unoccupied landscapes and their ability to cover the ground means that wolves can rapidly recolonize empty habitat. The potential for recovery is huge.

The countries in Western Europe that wolves have reclaimed were almost ideal for recolonization. They had abundant prey, little in the way of competition, and strong wildlife protections in place. Wolves flooded onto these landscapes. The pace of the return took many people by surprise, but it shouldn't have. The biology and ecology of this resurgence were largely predictable. But how wolves and people were going to deal with each other on tightly packed European landscapes was anybody's guess. Conservation still operated on the premise that nature and culture belonged in separate realms. Civilization and wildness in the form of wolves simply weren't supposed to mix. If the species was going to return successfully, a number of established mythologies would need to shatter. The European countries experiencing a wolf revival had at least one thing in their favor as they began this twenty-first-century cultural journey. They could look across the Atlantic and draw lessons from a wolf recovery already several decades in the making.

* * *

For close to forty years, Diane Boyd enjoyed a front-row seat on one of the world's most successful wolf recolonizations. She says this ranks her among the most fortunate wolf biologists alive. "In the bakery of life," she admits, "I got a really enormous cake." Recently retired after four decades studying wolves in the northern Rockies, Boyd still displays the hunger for wildlife that nourished her long career. She has shadowed the lives of scores of individual wolves. She followed their family dramas, witnessed their successes and misfortunes, and learned the difference between what kept them alive and what got them shot. Boyd probably knows as much about wolf recovery as anyone on earth.

Decades earlier, as an employee of a fledgling wildlife organization, the Wolf Ecology Project, Boyd tracked the first wolf to return to the Rocky Mountain states from Canada. The wolf, named Kishinena, moved back and forth across the international border, exploring new territory and consistently beating the odds. At one point, she raised seven pups alone after her mate died in a snare. In 1986, Kishinena's descendants moved permanently across the border and denned in Glacier National Park. The Magic Pack became the first pack to breed in America's northern Rockies in half a century.

Boyd watched the wolves from her Montana base near the Canadian border as they moved south. The remote cabin where she holed up north of Polebridge lacked electricity and water but gave her prime access to the wolves. The Wolf Ecology Project was a bare-bones operation, and Boyd learned to do everything on a shoestring. She tracked the wolves on skis, snowshoes, and sometimes a bicycle. She trapped them for measurements and blood samples, radio-collared them, and sent them on their way. She watched with admiration as they spread down the North Fork of the Flathead River on the park's eastern flank, skirted the cities of Whitefish and Kalispell, and padded through the farms and ranches of the Flathead Valley, eventually making it to Montana's Ninemile Valley more than a hundred miles south. She inspected their kills for clues about their habits, willing them to stay out of trouble. She marveled at the combination of luck and guile that kept them out of harm's way. The wolves needed more than luck to stay alive. In a culture not accustomed to having wolves around, the pioneering canids needed allies in order to survive.

In the Ninemile, a pack denned on the family ranch of brothers Ralph and Bruce Thisted. Their choice was a stroke of good fortune. Unlike many of their neighbors lower down the valley, the Thisteds were enthusiastic about the new arrivals. They were nearing retirement and no longer packed their land to the gunwales with cattle. When they learned about the wolves, they kept their livestock away from the den and removed as many of the temptations found on a ranch they could. Ralph would get up early to watch the pups through his binoculars. When the wolves' parents were both killed within a short time of each other, Ralph helped Mike Jimenez, a biologist with the U.S. Fish and Wildlife Service, keep the pups alive by sneaking road-killed deer onto their rendezvous site.

The Thisted brothers were fixtures at meetings at the Ninemile's small white-washed community center where locals met to discuss the wolves. There were strong opinions on all sides about the valley's new canine residents, but Ralph never became irate. "This isn't a battle," he insisted. "This is the way it should be." Everyone knew the Thisted brothers cared about the valley and the people who lived there. Their neighbors respected them. The fact the Thisteds were willing to accept the wolves carried weight. Rick Bass, a western author who chronicled the return of the Ninemile wolves, called the brothers "the Zen priests." He is convinced the wolves sought them out.

Ralph died in 2017, leaving his wife Betty with a stack of warm memories in the large timber-framed house in which she still lives. At nearly ninety, Betty remains active in land and conservation issues. She helps support educational programs for young women in the local schools. She is informed and strong-willed. When I met her at her house, she showed me a poster she had recently made, modeled on the periodic table of the elements, with each element symbolizing a significant event in her life.

Thirty-years after the wolves came back to the Ninemile, we sat on her deck with yellow-flowered balsamroots waving in the breeze below.

I asked why her husband welcomed the wolves back despite knowing they would kill the occasional calf and family pet. Her Mississippi accent softened the words she used to swaddle her late husband with affection.

"Ralph was a very quiet man," she told me. "He was not loquacious at all, until he needed to be." When he puzzled over what to do with the orphaned pups, Betty said she "could see a side of Ralph. . . ." She took a long pause, staring at the decking, and then looked up, changing the subject slightly: ". . . that's just what it should be all about as humans. . . . He was a beautiful man." The wolves were in luck. They had found people willing to rethink what counted as the right way to do things on a ranch.

The Thisteds persuaded a number of their neighbors to follow their lead. Montanans pride themselves on living where they can encounter big animals like bears, moose, and mountain lions. In the early 1990s, wolves were still a novelty in "the last best place." Down the road, there might be a hunting and trapping season. But for now, the Endangered Species Act meant the law was firmly on the wolves' side. The animals fared relatively well as they pushed farther and farther south from the forty-ninth parallel. Back in the Flathead, Boyd could hardly believe how well the wolf recovery was going.

Around the time the wolves arrived at the Thisted ranch, the U.S. government developed a wolf reintroduction project of its own. The 1973 Endangered Species Act required the return of wolves onto "unoccupied critical habitat." Yellowstone National Park, 350 miles to the southeast of Boyd's cabin, was the most obvious place for a government-sponsored return. The 22 million acres of the Greater Yellowstone Ecosystem is one of the largest intact temperate ecosystems in the world. It was chock full of the elk and deer wolves loved to eat. A second suitable site was identified in the wilderness of central Idaho.

The Magic Pack's arrival in Glacier predated the U.S. Fish and Wildlife Service restoration by almost a decade. Boyd wondered whether the government should risk infuriating ranchers and hunters by bringing wolves down to Yellowstone in trucks. People choose to live in the woods because they don't want to be pushed around. There were already seventy-five wolves south of the Canadian border, Boyd estimated, by the time Secretary of the Interior Bruce Babbitt carried the first set of reintroduced wolves into Yellowstone in 1995. Even today, Boyd thinks a great deal of anger would have been spared if wolves had made their own way back rather than being trucked into Yellowstone by the federal government.

Within two decades of the reintroduction, the biological questions about how best to ensure a successful wolf return were moot. The transplanted wolves had joined the naturally dispersing wolves to form a single interbreeding population big enough for federal protections to be lifted across Montana, Wyoming, and Idaho. The wolves could now survive without the heavy hand of federal law on their side. Responsibility for management returned to the states, which almost immediately reinstated hunting and trapping seasons.

The politics remains tricky, but the return of the wolf in the Northern Rockies has been a remarkable success from an ecological point of view. When I asked Boyd why she thinks it went so well, she cites a number of biological factors. Wolves, she says, are generalists, capable of adapting their diet and behavior to whatever conditions provide. "They live by their feet," she added, constantly on the hunt for conditions in which they can prosper. Even a settled pack spends eight to ten hours a day on the move. Wolves also have what biologists call *phenotypic plasticity*. This means their bodies change to adapt to different conditions. In the heat of the Middle East's Gaza Strip, wolves became smaller to manage body temperature more effectively in the hot and dry conditions. Ears are larger in southern latitudes, which makes them better at dissipating heat. In the Canadian Arctic, wolves grew big enough to take down caribou and to survive the long, cold stretches between successful hunts. Where prey is smaller, such as on Ellesmere Island, they shrank. The adaptable social structure of the pack is also critical. Pack numbers swell or contract depending on what conditions demand. Wolves can handle many of the ecological challenges that are thrown at them.

What Boyd witnessed in the northern Rockies forty years ago is now replaying itself in Europe, but with a couple of telling differences. First, no European country has intentionally reintroduced the wolf. They have made their own way back. It's a remarkable biological feat with significant societal implications. Wolves in Europe are not tainted by the stain of being reintroduced, something that still sours attitudes in America's Rockies. The second difference doesn't sound like an advantage. Europe has twice the people and half the land area of the United States. It has a longer industrial history, and more of its landscapes have been transformed to serve human needs. It lacks the extensive forests and remote valleys of Boyd's Northern Rockies. Wolves simply don't have the space in Europe they found in Montana and Idaho.

You might suspect this would spell disaster for a wide-ranging carnivore that subsists primarily on large ungulates. But European wolves turned out to have something different to teach from their cousins across the Atlantic. In the well-trampled lands of the old world, the wolf surprised everyone with just how close to people it could live.

2 THE WOLF NEXT DOOR

The old truffle hunter leaned on his spade and took a long drag on his cigarette. "*Duemila lupi*," he said. Two thousand wolves. He scanned our astonished faces and then looked down at the dirt and shook his head, not even trying to conceal his disgust.

We were standing in an Umbrian oak forest, having spent the last two hours watching the hunter and his dogs search for truffles. Two decades before, the regional authorities planted the rolling hillside not far from Perugia with oak saplings. When they put in the trees, they inoculated the soil with truffle spores. Now the woods produced a reliable harvest of the subterranean tubers each summer.

The black truffles the dogs had sniffed out that morning were not the region's most valuable treasure. A giant white truffle found in Tuscany a few years previously had fetched $330,000. But these black truffles provided Andrea, our host, with a modest income and a chance to be in the woods with his beloved short-haired pointers. The pointers seemed to enjoy it too. They earned an ear scratch and a bite of wiener for every truffle they sniffed out from under the duff. After Andrea had filled his bag with about forty nuggets of the black gold, the conversation drifted toward the area's wildlife.

Two thousand wolves was probably an overestimate for Umbria. The scientific consensus for the whole of Italy came in around there. But even if Andrea's guess was a little high, nobody denied Umbria had seen a dramatic resurgence of wolves. He showed us a photo on his phone of a wolf trotting casually through a wheat field close to where we stood. His point was well taken. There were tons of wolves making a living on this domesticated landscape. "You don't even want to raise this topic with the local shepherds," he warned.

The Italian gray wolf or Apennine wolf (*Canis lupus italicus*) is a smaller subspecies of wolf than the one repopulating the Northern Rockies (*Canis lupus occidentalis*). But like those wolves, it is a wily and opportunistic predator. In many parts of its European range, the wolf has found a way to occupy humanized landscapes. It slinks through the olive groves and vineyards that blanket Umbria's hills, making its home on a landscape more associated with wine, poppies, and the honey-colored stone of hilltop villages than with glaciated valleys full of grizzly bears. Italy has four-fifths of the land area of Montana, stuffed with sixty times the human population. Yet it has at least as many wolves and, if you trust Andrea's reckoning, quite a few more.

The Apennine wolf feeds mainly on wild boar and roe deer. When these larger prey items are scarce, it will eat rabbits, rodents, and berries. And, as Andrea was eager to remind us, it is not averse to snacking on the odd sheep. Most houses in the Umbrian countryside have a kitchen garden for fresh tomatoes, beans, and herbs for the pot. Many of these gardens have chickens. The chickens are always protected behind wire fences with a barking dog patrolling nearby.

The thriving population of wolves in Umbria represents quite a turnaround. Italian hunters and poachers followed the standard European playbook and depleted most large wildlife during the eighteenth and nineteenth centuries. By the end of the First World War, only a small population of wolves survived on the rugged flanks of the central Apennines. The Italian government created Abruzzo, Molise, and Lazio National Park in 1922 in the nick of time. The new park played a significant role in keeping wolves alive.

Wolf numbers remained precarious for another fifty years thanks to widespread poisoning and trapping. In 1971, the national government began a recovery plan called Operation Saint Francis named after the Italian monk said to have befriended a wolf. New laws prohibited hunting wolves or killing them with poisoned bait. In 1976, wolves became a fully protected species in Italy, and their numbers quickly took off. A string of Apennine parks created along the spine of Italy in the 1980s helped the wolf population surge. Wolves spread north from their Abruzzo stronghold toward the Alps and west toward Rome. By the turn of the twentieth century, dispersing Apennine wolves had passed through the Alps and into Switzerland and France. They surprised just about everyone by their tolerance of humans. One wolf was killed by a truck on the edge of Turin. Another pack denned in the Castel di Guido park under the flight path of Rome's Leonardo Da Vinci International Airport. Skiers at an Italian resort were surprised one winter to see a pack passing through the village. Wolves were not only back. They were back and making a living right under people's noses.

* * *

You don't have to travel far in western Europe to find people angry at the wolf's success. Those accustomed to working the land without wolves feel their livelihoods threatened. In addition to the irate shepherds, Italian hunters now have competition for the wild boar (or *cinghiale*) prized in local cuisine. Over the Alps, French farmers forced to deal with wolves are fuming. Wolves, they say, make ten thousand attacks on livestock each year. To make matters worse, these aren't "their" wolves. They are interlopers from Italy. Unhappy herdsmen regularly bring their sheep to clog up the center of provincial towns to draw attention to the crisis. The compensation offered by the government, they insist, is not enough. They are overrun, they claim, pointing out that wolf numbers reached the government target of five hundred several years ahead of schedule.

Italian, French, and Belgian farmers are not the only ones feeling the pinch. Large populations of wolves that survived the Cold War in

Romanian dictator Nicolae Ceausescu's hunting preserves spilled out of the Carpathian Mountains, buttressing wolf populations in the Ukraine and Hungary. Recovering Polish wolves fanned out into Slovakia, Czechia, and Germany. From there, wolves dispersed west into the Netherlands and north into Denmark. Wolves from Russia replenished the Scandinavian populations as fast as intolerant farmers could kill them. Wolves from Slovenia's Dinaric Alps bred with wolves from northern Italy, rejoining a pair of subspecies separated for two centuries. The Iberian wolf population in northwest Spain and Portugal is also increasing, thanks to the abandonment of marginal farmland on the peninsula.

The fact there are pockets of hesitancy about the resurgence of wolves makes perfect sense. Wolves do kill sheep. And let's remember, these wolves are top-level predators. We are not talking about a returning vole or a recovering songbird. They are eighty-five-pound carnivores with jaws that can crack the femur of a large deer. It is not unheard of for wolves to enter a frenzy and practice surplus killing, especially when they find themselves in the midst of domestic livestock that have lost their fear of predators. In North America, a pack of three adults and five pups near Dillon, Montana, massacred 120 sheep in one night. Wolves in Wyoming slaughtered nineteen elk gathered at a winter feedlot in a single incident, far more than they could eat. And while there are vanishingly few cases of wolves attacking humans, the primal fear remains strong. When Mexican gray wolves were reintroduced to Arizona and New Mexico, local citizens built cages along the roadside to keep children safe while they waited for school buses.

It is tempting to think the wolf resurgence in Europe will deteriorate into a replay of familiar battles. Farmers against city-dwellers. Conservatives against progressives. Nature's interests against the interests of those who work the land. These fault lines are certainly evident. But they are not the whole story. It is no longer the 1850s. Beneath the predictable chatter, something different has started to emerge. And this difference is key to opening up the prospects for the recovery of wolves—and the recovery of other wildlife—across the industrialized world.

* * *

Nineteenth-century Europeans struggling against hunger and disease needed a clear partition between the human world and the wild world that threatened them. The mental separation they created helped solidify the sense of their humanity. Thinkers associated with the Enlightenment asserted that reason cut a decisive line between humans and everything else. The scientific and industrial revolutions provided all the evidence they needed for this claim.

Romanticism pushed back against the authority of reason, but the environmental awakening it spawned unwittingly solidified the gulf. The model of conservation that emerged after romanticism crossed the Atlantic declared that wildlands were at their peak only in the absence of people. This model had consequences. Miwok and Paiute Indians were forced out of Yosemite to satisfy the fantasies of conservation icons like John Muir and Theodore Roosevelt. Nearly a century after the first parks, the 1964 Wilderness Act declared wild areas were places people can visit but not remain.

Wildlife and especially wild predators are the emblems of this spatial division. The roots of the word *wilderness* lie in the Old English *wildeor*, which means "wild beast." The place for wild beasts was absolutely not the place for people. And, in turn, the places people lived were off limits to wildlife. No form of wildlife was less welcome in civilized spaces than the wolf. "Killing wolves became a symbolic act," writes Lopez, "a way to lash out at that enormous, inchoate obstacle: wilderness." What could be more emblematic as a justification than the fact wolves, it was said, hated music? The strict segregation between people and nature imposed by this type of thinking was highly offensive to Indigenous people. The idea humans might exist on a different metaphysical plane from their animate surroundings made no sense at all to an Indigenous mind. But early environmentalism was blind to its racial underpinnings. For most of the twentieth century, U.S. and western European ideals steered the conversation about wildlife with little social conscience.

At long last, something in those ideals is changing. Whether environmentalists are more alert to the injustices perpetrated on Indigenous people or whether they are experiencing the malaise of the industrial life George Monbiot calls *ecological boredom*, the entrenched views are starting to crumble. It is more common to find people wanting to have wildlife around. It makes them feel more human and their lives more vital. The Umbrian truffle hunter admitted his respect for wolves. When he showed us the photo of the wolf on his phone, he was beaming. He's a hunter, and he admires a tough opponent. He wants a landscape full of wildlife, just like they do in Montana's Ninemile Valley.

Greater awareness of ecology has also shifted the story around predators. Prey populations benefit from being hunted by wolves. Deer herds are less likely to overgraze or incubate disease. Herbivores become more wary. The effects ripple down through the ecosystem. When grazers are more skittish, streamside vegetation recovers. Creeks contain less sediment. Insects and fish flourish. Water temperatures cool. The ecological benefits cascade down from the top. The facts accumulating around wolves today are not the same facts as circulated in the nineteenth century. Many of the old myths have been exposed.

How effectively this new attitude will stick depends a lot on what happens in places like the Netherlands. If large carnivores require separation from humans on Yellowstone-sized landscapes, the wolves' return will be short-lived. Guillaume Chapron, a wildlife biologist at the Swedish University of Agricultural Sciences, thinks a coexistence model is starting to replace a separation model. A softening of the strict ideological boundaries splitting people and wildlife is underway.

There are some striking examples of the coexistence model at work. The International Union for the Conservation of Nature studied how Indian villagers in Gujarat live alongside crocodiles. The villagers build islands in the lakes to give the crocodiles a safe place to bask. Fisherman move their boats into ponds the night before they fish, so the crocodiles know to stay out of the way. Villagers and crocodiles both seem to have learned the drill. Maasai tribespeople in Kenya's southern Rift

Valley coexist with lions with very little conflict. The lions have adapted to the Maasai's seasonal movements by changing where they look for prey at different times of the year. Tigers in Nepal's Chitwan National Park hunt when people are not around to avoid dangerous encounters on forest paths.

Katmai National Park in Alaska contains a different example of coexistence between humans and large carnivores, one startling to witness firsthand. Tourists flock to Katmai's Brooks Falls to see brown bears dining on sockeye salmon. Large sow bears with cubs hurry a few feet past the kitchens at Brooks Camp. The proximity is unnerving for those schooled in the idea of bears as irredeemably wild. But there is plenty of fish for the bears to eat, and they don't seem to mind encountering a few humans on the way. Cooks and college kids working summer jobs lean out of doors and bang pots and pans to remind the bears to keep moving. Clusters of visitors wait patiently with rangers as bears take naps on footbridges. Bad encounters between people and bears at Brooks Camp are virtually nonexistent. "These aren't normal bears," said the park ranger during the mandatory safety briefing I attended. "These are Brooks Camp bears." People and bears at Brooks Camp have learned an etiquette for coping with each other's company.

The idea of an etiquette for coexistence may sound fanciful. But it reflects a deeply embedded truth. Wildlife and humans are evolutionary kin. We evolved alongside predatory animals and learned a thing or two in the process. It is true that coexistence with large carnivores is not simple. But nor is it impossible. Buried somewhere in each party's genes is an ancient wisdom about how to make space for each other. For industrialized cultures, this can feel deeply alien. It can feel especially alien when dealing with animals powerful enough to kill you. Developing or redeveloping these etiquettes requires creativity and practice. It won't happen without a willingness to invest the money to create the conditions for change. It also requires another sort of investment—an investment in a whole new attitude about the wild.

* * *

In the months following the discovery of the Flevoland wolf, Dutch ecologists and environmentalists worked frantically to smooth the wolf's return. The Dutch had particular interest in doing it right. The Netherlands is the headquarters of European rewilding, the movement to bring wild animals and ecological processes back onto suitable portions of the landscape. They needed to lead by example. And if the Dutch can do it on one of the most highly developed and densely populated landscapes on earth, anyone can.

The Netherlands provides significant challenges for wolves. It has lots of people, an intensive agricultural sector, and a cultural tendency toward order and control. Willem van Toorn, a well-known Dutch poet and public intellectual, was vocally opposed to rewilding and the return of wolves. He insisted Dutch landscapes derived their meaning from their *human* history and not their natural history, something wolves were bound to disrupt. He thought the human history on Dutch landscapes should have priority. Everything in the Netherlands at this point, he said, is culture.

Van Toorn's voice, however, was not the only one in the debate. Educated professionals bumping elbows with college students in Amsterdam's cafes toasted the wolf as an exciting addition to the landscape. The Dutch consider themselves a progressive, environmentally minded people. To many, Van Toorn represented the old school. Wolf advocates found themselves with thousands of allies. Children sat transfixed in classrooms as they learned about the animals' return. Biologists poured over maps looking for suitable habitat. The Dutch also did not face this challenge alone. They had neighbors with more than a decade of experience of their own wolf recovery. With similar landscapes and a variety of overlapping values, Germany offered a model for what lay ahead.

The wolf population in Germany had rocketed from zero to more than 350 in less than two decades. The collapse of the Soviet Union created an unexpected opportunity for wildlife. The departure of NATO and Soviet forces turned dozens of military bases into stepping-stones for wolves. Researchers found wolves preferred the forested bases to protected natural areas. The land was off-limits to anyone not in uniform.

Poachers avoided it. Birdwatchers stayed away. As long as the wolves learned how to dodge the occasional tank, they found plenty of deer and enough undisturbed cover to successfully raise their pups.

When they started dispersing from the bases, most of the packs headed to the less populated parts of Germany to the north, though it wasn't long before sightings occurred in every German state. Wolves at one point were increasing by a third each year. Attacks on farm animals remained low. There was abundant prey and not many other wolves to provide competition. But as the German landscape filled up with wolves, livestock deaths began to mount. Keeping everyone happy was going to require some creative peace making. And right on cue, some creative peacemakers appeared.

* * *

Nathalie Soethe looks as much like a German wolf enthusiast as any person can look. I first encountered her in the Netherlands at a symposium on rewilding. She is a thoughtful and pragmatic presenter, laying out the challenges clearly and detailing the policies most likely to work. But the Germanic formality evaporates as soon as the conversation moves to her own work with wolves. Her face beams and her brown hair bounces as she talks animatedly about her efforts to assist the carnivore's return.

Soethe's day job is in the department of landscape economics at the University of Greifswald on Germany's Baltic coast. She works on sustainable agriculture. Despite the academic credentials, Soethe is no stuffy university type. She is as comfortable doing hard physical labor as she is pouring over spreadsheets on sustainable land use. She laughs easily at herself and focuses on how to get things done. You know within five minutes this is someone who would have no trouble maintaining a friendly conversation with a grumpy farmer, should the situation require. It's a trait that has served her well.

Soethe grew up in a village outside Hamburg. She remembers the outsized role the local farm played in her childhood. The sound of tractors was the background score for her weekends. She and her friends threw mud on each other in the fields and gazed wide-eyed at newborn lambs. When choosing what to study at university, Soethe bounced back and forth between agriculture and wildlife. Growing food and working the soil felt like essential human activities. At the same time, she loved the surprise of a fox trotting along a hedgerow or a deer darting across a ploughed field. She chose agriculture and headed to Christian Albrechts University in Kiel for her degree.

On graduation, with her agronomy credentials in hand, Soethe went to South America to do research in forest ecology. She met farmers in Brazil and Ecuador living alongside dangerous animals like crocodiles and anacondas. They were poor, and their options were limited. Either a crocodile took their animals, or the farmer killed the croc. Soethe noticed that when the locals took matters into their own hands, environmentalists from Germany and other developed countries howled in protest. It did not seem fair.

Soon after Soethe returned from South America, the first brown bear in 170 years showed up in Germany. The bear had struck north from a small population restored to Italy's Trentino Alps. As the bear lumbered toward the German border across a thin sliver of Austria, the Bavarian Minister of Environment put out a statement celebrating its imminent arrival and promising the German people would be good hosts. Hikers spotted the bear in the middle of May shortly after he entered Germany. The press named him Bruno after the Italian term for brown bear—*orso bruno*.

The German public quickly fell in love with the crafty ursine, following his every move. They were appalled when, after a two-month display of somewhat antisocial behavior—involving the killing of sheep, chickens, and even a hamster that Bruno allegedly ate while sitting in front of a Bavarian police station—the Ministry of Environment declared him *ursus non grata* and put out a shoot-to-kill order. A hunter executed Bruno the same day. Fresh from her experiences in the Amazon, Soethe was shocked at how a country as rich as Germany with so many options for putting food on the table couldn't keep a single bear alive for more than a few weeks. She wondered what a more forgiving relationship with wildlife required. If large animals were to survive on densely populated landscapes such as Germany's, people and wildlife would have to figure out ways to get along.

A few years after wolves showed up in Germany, Soethe found herself at a workshop on landscape management listening to a lecture by a member of the organization WikiWoods. This loosely coordinated group of volunteers gave up their weekends to help Germany's forests by planting and maintaining trees. Soethe approached the speaker at the end of the talk and asked what he thought about a similar organization for wolves.

Soethe created WikiWolves later that year, fueled by the idea that a group of well-intentioned volunteers could make sharing the landscape with wolves much easier for farmers. Germany, like the Netherlands, was required to protect wolves by EU law. The German government had agreed to provide compensation for livestock lost to wolves. But to access the compensation funds, farmers needed to show they had fully implemented deterrence measures on their land. Building fences and organizing guard dogs took time, something Soethe knew was in short supply on a farm.

WikiWolves became an organizational hub for people willing to donate their time to farmers. This meant long hours digging post holes and putting up electric fences. The Danes had developed an electric fence effective at keeping wolves away from sheep, but it had to be built properly. The strands had to be at the right height, and the posts tight enough to prevent sag. The wires needed to be clear of anything that would create a short circuit. After learning how an electric shock feels, wolves quickly gave the fenced area a wide berth. Rarely would wolves risk jumping the fence, and, if they mastered that trick, wildlife agencies were allowed to kill them.

Soethe recalls arriving nervously on a farm with a handful of volunteers on WikiWolves' first outing. She found the farmer pacing around, uncomfortable at the prospect of spending the day with a group of city-dwelling wolf lovers. The farmer's neighbor came over with a scowl on his face to make it clear he had his eye on them. The farmer's wife promised them lunch, and the volunteers set to work putting up fences on two sides of the field. As the six WikiWolves volunteers labored with post-hole diggers and sledgehammers, the neighbor showed up several times on the edge of the field in his car. At the end of the day, he came over, slipped his phone number to Soethe, and asked her to get in touch. He had been won over by what he saw. A year later, the same farmer stood up at a convention hosted by a conservative political party designed to demonize wolves and said "WikiWolves. . . . I know them and they are top people."

It quickly became clear to Soethe that the social labor performed by WikiWolves was every bit as important as the physical. The organization put those who didn't like the idea of wolves into close contact with those who did. Farmers and wolf advocates worked shoulder to shoulder for long hours. They opened up conversations and heard each other's point of view. By talking about their lives and challenges, they started to diffuse some tension. Volunteers who cared about wolves also tended to care about farm animals. They wanted both of them protected. By giving up their weekends, they showed a willingness put skin in the game. The farmers saw firsthand the enjoyment WikiWolves volunteers got from working the land. The wolf advocates gained an appreciation for the farmers' work in maintaining open space.

Wolf enthusiasts also stepped up to help farmers in France. A group called Pastoraloup arranged for volunteers to camp in the mountains at night to help shepherds keep their sheep safe. It lessened the number of animals taken by wolves and decreased miscarriages among pregnant ewes. The French Ministry of Environment and the Worldwide Fund for Nature pay for the volunteers' training. The farmers appreciate the benefit they receive. They also value the sense they are not fighting the battle alone.

Soethe is philosophical about the path ahead. "I worked as an ecologist, a researcher, in wild nature. But at the same time, I love agriculture," she said. "For me, both are very important. I want to build bridges." Landscapes have an ecological carrying capacity for predators. But they also have a cultural carrying capacity. The ecological capacity is determined by the prey base and the habitat. The cultural capacity is determined by the social acceptance the animal manages to achieve. Soethe's goal is to move the cultural capacity as close to the ecological capacity as she can. "Europe is very densely populated," she admits. "But we have money. The density is no argument for me." Germans and other Europeans simply have to make the investment in figuring out how to share the landscape with wolves.

Taking their lead from Germany, the Dutch became proactive in their efforts. Provinces around the Veluwe offered money to farmers for the preventative measures seen to work in Germany. They funded 100 percent of the costs of approved techniques in at-risk provinces, keen to put the measures in place before more wolves arrived. If the wolves never developed the taste for sheep, it was more likely they would stick to their natural prey.

Farmers had two years to take authorities up on the funding. After that, there would be fines for livestock left unprotected. Harmonious coexistence with wolves would require good behavior from both sides. Jansman, the forensics expert, is quick to agree with Soethe about where the real work lay. "Wildlife management," he says, "is really an application of people management." And this is not only about what people do. It's about what they think. In other words, the Dutch didn't just need a new set of habits about how to farm. They needed a new way to conceive of the landscape and their proper relationship to it. If Willem van Toorn and his followers were to be convinced, they needed to find something to replace the gulf that had built up in their minds between culture and the wild. For that, they needed someone to bring to the table a better *philosophy*.

3 SHAKESPEAREAN

Martin Drenthen's face has an impish look not typically found on a university professor. Drenthen teaches philosophy at Radboud University in Nijmegen on the eastern edge of the Netherlands. He and the graduate students he brings to Radboud are officially part of a science faculty. The concrete and glass building where they work smells more of chemistry experiments and engineering models than of the coffee and cigarettes associated with people who argue over Martin Heidegger and Maurice Merleau-Ponty. The glint in Drenthen's eye may stem in part from how he likes his role as an interloper. The boundaries he and his students push from their philosophy offices have the potential to be as transformative of Dutch culture as anything his engineering colleagues might design.

Nijmegen was founded in Roman times and is the oldest continually inhabited city in the Netherlands. It occupies a strategic location on the River Waal. The bridges spanning the wide and flat river were the targets of Operation Market Garden, the largest airborne assault by allied forces during the Second World War. British and American bombs demolished sections of the city during the landing. A museum to the battle on the edge of town attracts visitors from all over the world.

The town center is an awkward mix of seventeenth-century Dutch renaissance style and postwar brick and cement. Its clash of styles and long history of confrontation makes it fitting that Nijmegen is the epicenter of European attempts to rethink how humans and nature interact. Rewilding Europe has its headquarters here. Nearby Wageningen University has a special professorship in rewilding ecology. At the Millingerwaard Nature Reserve on the edge of town, summer dikes have been taken out to let seasonal floods shape the land again. Nijmegen is driving a change in Dutch thinking about how to live alongside the wild.

Drenthen has become one of the public faces of this change, appearing frequently in newspapers and on talk shows to discuss the wolf's return. He lacks the mane of curly brown hair worn by his friend Hugh Jansman, but Drenthen's face also creases with a massive grin when he parleys about the wolf. An animated talker, his large hands and elbows whip the air into a frenzy when driving home his points. And right now, his points have become particularly urgent. The wolf's arrival in the Netherlands has given him the ultimate case study for his work.

* * *

Drenthen, like Jansman, had been eagerly awaiting the arrival of the first wolf in the Netherlands. Around the time it appeared, Drenthen and his partner, Saskia van Boven, spent several months in western Montana talking to wolf and bear experts to develop a feel for living among large predators. Setting up shop in the college town of Missoula, they had front-row access to numerous wildlife experts, land managers, and environmental philosophers. They set out to learn what they could from people much more accustomed to being around carnivores than their friends back home.

Drenthen and van Boven reveled in the proximity of the wild. They could step out the front door of the home they rented near the railroad tracks and see elk wintering on the flanks of Mount Jumbo less than a mile away. Part of the point of coming to Montana was to be out on the land as much as they could. They paid close attention when offered a lesson on how to use bear spray. With a hand on his polished head, Drenthen cracked himself up at the idea of entering the woods with a can of capsicum to deter charging bears. The Dutch are not used to thinking about taking precautions when going out for a walk.

The couple took a number of trips to Yellowstone and Glacier National Parks to experience wild carnivores. In Yellowstone, they watched wolves in the Lamar Valley with Rick McIntyre. McIntyre is an award-winning author and a Yellowstone legend. He has spent over eight thousand days studying the behavior of the park's wolves, logging more than a hundred thousand sightings since their reintroduction in 1995. At one point, he did not miss a single day in the field for fifteen years. Nobody has spent more time watching the dynamics of wolf packs through a spotting scope.

Drenthen and van Boven were briefed in advance about what to expect. McIntyre arrives in the valley at dawn, however frigid the weather, wearing his trademark jeans, patched-up jacket, and wool hat. He starts scoping out the wolves' typical haunts. When he locates a pack, he plants his tripod, pulls on his gloves, and starts to observe. Whenever a wolf gets up, yawns, or moseys over to an elk carcass for a few bites, McIntyre notes it in his voice recorder. He recognizes most of the Lamar's wolves by sight and is familiar with their personalities. He talks about them as if he were talking about an admired athlete or a favorite friend, not a member of another species. Every wolf is unique—so unique, in fact, that he has written award-winning books about three of them. McIntyre has come to think of wolves in Yellowstone as living out a Shakespearean drama, animated by competing temperaments and passions. He wouldn't have had much truck with Descartes.

Drenthen absorbed McIntyre's teachings about wolves over several days in the Lamar. He and van Boven had dinner at McIntyre's cabin outside the northeast entrance of the park. With snow piled high outside, they heard tales of the wolves who had shaped the Yellowstone story. McIntyre impressed on Drenthen the importance of seeing animals as individuals, each with their own strengths and idiosyncrasies. If you look at them this way, the gap between the human and the wild world does not appear so vast. The kinship feels real.

Drenthen knew this very personalized approach to wildlife would take some learning by the Dutch. It went against years of scientific teaching in wildlife biology that discouraged humanizing—or anthropomorphizing—the wildlife being studied. Farmers had already created the narrative of wolves as ruthless killing machines. The wider Dutch population, having not interacted with wolves for more than a century, had little basis on which to argue. Changing the well-entrenched perception of wolves was going to be tough. But Drenthen had at least one consideration working in his favor. The wolves in the low countries could still be counted on the fingers of two hands. Like Boyd's experience with Kishinena and the Thisted brothers' experience with the Ninemile pups, this was still a personal story. The handful of wolves in the Netherlands were identifiable individuals, each one distinctive and each one starting with a fresh slate. This newness created an opening.

* * *

The landscapes Dutch wolves entered contained far more people than Montana. But the habitat was good, and the prey was off the charts. Wolves had arrived at a time of extraordinary growth in their natural foods. Farmland was teeming with the species they liked to eat.

Roe deer were almost extinct in the Netherlands in the 1870s. By 1930, their number had crept up to between three and four thousand. Protection by the government and a healthy regrowth of woodland have brought the numbers to over a hundred thousand today. Across Europe as a whole, there are close to ten million roe deer. Red deer, a species closely related to North American elk, tripled in the Netherlands in the last half century. In France, they quintupled. There are now at least 2.5 million of them in Europe as a whole.

Across the channel in the UK, there are more deer today than there have been for a thousand years. I can confirm that anecdotally. As a child, I never saw deer—dead or alive—anywhere in the region I grew up. The last time I was in southern England, I saw several, some hit by cars, others grazing in woodland, still others watching from the edges of fields. On a taxi ride one night, the driver told me he wouldn't use the country roads after dark for fear of colliding with a deer. The countryside is now home to a large mammal I simply did not see in my youth. Back then, you had to go to Scotland to see a deer. Children growing up across wide swathes of Europe today have a different baseline for ungulates from anyone over age fifty. The recovery has been almost too successful. Overgrazing is a problem. Governments sign contracts with deer hunters to keep crop damage in check. This is not good news for farmers, departments of agriculture, and late-night drivers. But it is definitely good news for the wolf.

Wild boar numbers are also booming. The European Union estimates the native boar population is over ten million. Warmer winters, more productive farms, and the boar's willingness to eat whatever is available have all worked in its favor. The boar are also prolific reproducers. Females reach breeding age within a year and can produce two farrows of piglets every season. They are happy to root around for whatever they find in a farmer's field or a city park. They compete with truffle hunters like Andrea for his livelihood. The Dutch shoot thousands of wild boar each year to control their numbers. In the UK, they are accorded the same legal status as rabbits, pigeons, and gray squirrels. This means they are a nuisance animal and can be shot without a license year-round. Boar may be a growing bother to farmers across Europe. But they are a tasty snack for a wolf.

The profusion of available prey has the potential to shape the current debate about wolves. If there is enough wildlife to eat and enough effort invested in deterrence, wolves are less likely to trouble sheep or cows. Their predation of deer and boar will help reduce the cost of crop damage for state agencies. If farmers know they will be fairly compensated for their losses—whether these are sheep that disappear or cattle that don't put on as much weight—the economic reasons to vilify wolves diminish. The potential for coexistence emerges, nourished by organizations like WikiWolves and Pastoraloup. If wolves are allowed to stay, rural ecologies become more dynamic. I know from growing up in England what it is like when the biggest animal you expect to see is a fox or a rabbit. Nature is cute but rarely breathtaking. Diverse predator-prey relationships are the lifeblood of natural systems, and a dynamic ecology pumps energy into cultural life. "The wolf," said Lopez, ". . . takes your stare and turns it back on you." Wolves adds complexity to human experience. And with the complexity comes depth. A landscape is no longer just scenery when you know wolves are around.

* * *

A year after his Montana visit, I caught up with Martin Drenthen to ask how things were going with the wolf in the Netherlands. He thinks an uneasy cultural détente is in place. The Dutch are happy to be participants in the wolf renaissance across Europe. There are now three packs in the Veluwe, one of which has grown to ten wolves. They don't need go far from the forest to find prey. Livestock impacts are minimal. The opportunity to see a wolf remains a novelty for the vast majority of the Dutch. The situation at the moment is about as ideal as it could be. But Drenthen is uncertain how long it will last.

"I'm not optimistic," he says with a sigh when I ask him to predict the next few years. He knows more livestock are going to be killed. He expects the costs to increase and the agricultural community to feel victimized. He wonders how willing the Dutch will be to accept restrictions to protect the new arrival. Having lived near the borders of Germany, the Netherlands, and Belgium all his life, Drenthen recognizes differences in national character. The Dutch, he says, like a well-ordered, predictable landscape. The wolf threatens that.

As soon as the words are out of his mouth, he cracks a smile and waggles an optimistic finger in the air. "But I do hold out a little bit of hope," he says. "The Dutch have a strong sense of justice. It is no coincidence the International Court of Criminal Justice is located in the Hague." Drenthen thinks his fellow citizens are prepared to stand up for what is right. It is not impossible, he thinks, that the Dutch will see the injustice of expecting African farmers to live with elephants and Indian villagers to share the forest with tigers if they themselves are unwilling to put up with wolves. Like Nathalie Soethe in Germany, Drenthen thinks there are basic questions of fairness at stake. People in the Netherlands can't treat themselves differently from people in the Amazon. The Dutch, he suspects, will understand this.

Drenthen also thinks a deeper ethical philosophy is taking hold around animals. What is right is no longer just a question of what is right for people. Many of his countrymen believe moral regard is owed to the wolf itself, a species eradicated without mercy a century and a half earlier. Such a commitment nudges ethics beyond the human sphere toward nonhumans. This concern for other species is evident in groups like Wolven in Nederlands. It also lurks in dry pieces of legislation like the EU Habitats Directive, which mandates the reintroduction of native species to all European lands from which they are absent. There is a growing recognition in many modern European societies that the world is not just for us. It is for nonhumans too.

Having a conversation about wolves with Drenthen takes you on a rollercoaster ride between optimism and hard-nosed pragmatism. He is convinced things are changing, but he knows the fault lines run deep. When we speculated about how to ensure a successful return of wolves, Drenthen laid out the leap he thinks industrialized people must take. "The main argument that I make is there is a dualism still playing in all discussions of nature conservation," he said. The placing of culture and nature on two ends of a spectrum is a problem. "That very quickly gets to the idea that protecting nature means having these reservations," he said. The borders between the civilized and the wild are treated as impermeable.

Muir and Roosevelt's legacy, in other words, the one cemented in the U.S. Wilderness Act, creates a barrier to the kind of thinking that would support modern European wolves. The idea that wildernesses are reserved for the *wildeor* (wild beasts) and that people can live only in protective isolation from them is a damaging fiction.

"That has never been true," Drenthen said, leaning forward in his chair. "Some animals are perfectly able to cross those borders all the time. Birds, for instance . . . and everybody likes to have birds in their garden. And a wolf, because it is so mobile and so smart, could not care less about that border between nature and culture. And that's why I think the wolf is such a rewarding example to talk about." The wolf brings into focus the need to rethink an outdated notion of separation, one first bequeathed to the Dutch four centuries ago by Descartes.

Wolves, perhaps more than any other wildlife species, have literally crossed this border. Humans have a fifteen-thousand-year history of adapting to wolves during the process that created the domestic dog. Being in their company—first, as they prowled around the edges of campfires sniffing for scraps and, eventually, as they curled up in front of the hearth with their heads on our feet—should have done something for how humans thought about the wolf. They went from villainous beasts to loyal companions. The civilized and the wild broke free from their conceptual cages and found common ground somewhere in-between. Yet this change in fact never resulted in a change in mind, as people continued to villainize the wolf. The hope this can end is what motivates those who advocate for a different relationship with wolves.

"It is easy to say we are not used to living with wildlife in the Netherlands," said Drenthen. "But of course we are. Rats, mosquitoes, seagulls. We have ways of dealing with them." Many species of wildlife adapt to living alongside us without us even knowing it. Deer know how to find the lushest grass on farms and in the suburbs. Rats and mice sniff out the leftovers from our tables and supermarkets. Bees and beetles make a home in our gardens. What sort of cohabitation can emerge with the wolf?

Life in the Netherlands in the 2020s, Drenthen points out, is not as precarious as the 1850s. "The differences are important. We are more resilient, and we can make ourselves much more resilient. We don't have to shut out nature as an enemy. We can afford to leave room for those other beings." Drenthen promotes the idea of a "nature-inclusive" society.

It is a bold agenda and one made plausible by what we have come to learn about living with wildlife. Animals are smarter and more adaptable than we imagined even a decade ago. "It is not the numbers that are the problem. It is the behavior," said Drenthen. What matters is creating desirable interactions. This requires an updated understanding of what you are dealing with. "You have to take the animals seriously as beings with agency who have some kind of justified claim on space." They don't just have agency; they also have culture. Drenthen thinks today's wolves are not the same as wolves living 150 years ago in the Netherlands. They have learned and continue to adapt to being around people. The landscape they occupy has changed. A twenty-first-century wolf may have a biology largely similar to that of a nineteenth-century one, but it has a different set of smarts.

Taking the agency and culture of wolves seriously is to take what Rick McIntyre taught Drenthen to heart. Animals are individuals, each with the capacity for learning. "We should make sure these wolves stay shy enough. Once they feel they have nothing to fear, then we are bound for trouble. We should start trusting other beings and developing different modes of trust—trust between different modes of society but also trust in animals not wanting conflict with humans." Agency, trust, learning. These are ways of thinking about wildlife often missing from industrialized societies. But Drenthen is convinced it is time for them to come back.

None of these shifts stands a chance of happening, Drenthen warns, unless concerted efforts are made to bridge the rural-urban divide that still fractures Dutch society. To create modes of trust that reach beyond the human species, people need to do a better job of ensuring that trust exists within it. In rural economies, Drenthen points out, people tend to be more aware of their vulnerability, more conscious that nature can be a threat. A single hailstorm can destroy a crop. A late freeze can ruin a growing season before it begins. In 2019, 38 percent of Dutch farmers lived below the poverty line. The prospect of losing a couple of sheep to a wolf is worrying. This vulnerability is not felt as acutely by urban people. In the cities where most Dutch live, modern infrastructure and services insulate people from most surprises. Urban dwellers need to recognize the reality of being vulnerable to natural cycles. They need to sympathize with those in the rural economy. This is why Drenthen admires Soethe's work with WikiWolves so much.

At the same time, farmers need to think more sympathetically about life in the city. "It gets awfully lonely," Drenthen says, "if you only see people and no other organisms." In cities, the desire to know there are big animals roaming the landscape is stronger than it is for those with sheep to protect. Honoring this desire may feel unfair if you bear the costs of living with wolves, but the desire for the return of predators needs respecting. If the economic costs created by wolves are shared more equitably, some compromises should be possible.

These changes will take a while. But for those who envision a different future with wildlife, they are worth it. "The Netherlands is a rich country," Drenthen points out. There are obligations stemming from that wealth. "Wolves," he suggests, "are much easier to live with than elephants." Drenthen holds out hope his fellow citizens can appreciate that and will play the role in predator recovery they can afford.

He had better be right. A few weeks after we spoke, a wolf showed up in Noord Brabant, a southern Dutch province where farmers were not yet ready for its arrival. It killed more than fifty sheep in a week. A farmer shared a video of the wolf harassing a ewe before trotting off with a lamb in its mouth. The agricultural community was outraged. The old prejudices reared their heads. An organization called Wolf-Fencing Netherlands immediately stepped up in the province to provide the same service as WikiWolves in Germany.

* * *

Sitting in her Montana cabin 4,500 miles away from Drenthen's office, Diane Boyd is confident about the future of the animal she has devoted her life to studying. "Wolves will always be here," Boyd predicts. They survived the extinction of the megafauna during the Pleistocene. They made it through the wildlife exterminations of the eighteenth and nineteenth centuries. When Boyd started her work documenting their recovery four decades ago, she never thought wolves would be as successful as they have been. They were too big, required too much prey, and aroused too many emotions to flourish on twenty-first-century landscapes. Unlike foxes or coyotes, she thought, wolf packs could not survive at the bottom of someone's garden or in a suburban park. Even in thinly populated places like Montana, they still face threats, Boyd says. There are still those who prefer "shooting 'em, hunting 'em, and hating 'em."

But these big canids have shown Boyd how remarkable they are. They are highly resilient survivors. Their intelligence and adaptability have ensured a recovery beyond anything she thought possible in the late 1970s. Wolves flourish from Africa to the Arctic. The fact they live by their feet helps. If pushed, wolves can simply trot their way out of trouble. The social environment in which they find themselves continues to change. State-sanctioned poisoning of wolves has become rare. Public sentiment has mostly swung in their favor. This all makes a repeat of their earlier eradication highly unlikely. She doubts that even the challenges of climate change will spell the end of the wolf. The wolf, Boyd is convinced, is here to stay.

* * *

After wolves showed up in the Netherlands, there was only one country left in continental Europe where *Canis lupus* had yet to return. Belgium's population density is comparable to that of the Netherlands, with only three-quarters of the land area. It has a similar history of industry and commerce. Ninety-seven percent of the Belgian population live in cities, which are known more for their artisanal beers and chocolates than for their wildlife.

Belgium shares its longest border with France, but it also butts up against the Netherlands, Germany, and Luxembourg. Two decades into the twenty-first century, all these countries had wolves. It was only a matter of time before Belgium joined the rest of the continent and completed the wolf's European return.

The Belgians have always tended to be a little more free-wheeling than the Dutch when faced with change. Philosopher Jan Opsomer told me his country's location in the battlegrounds of Europe had left its mark, particularly in the north. "There is a rebellious nature to the Flemish," he told me, "more Romantic than Germanic, unlike our neighbors in the Netherlands." Belgians tend to admire the rugged individual and are ambivalent toward the established order. These national traits were about to prove helpful.

In January 2018, a wolf wearing a radio collar left a pack near Berlin and headed west. She crossed the border into the Netherlands and veered south. The wolf wandered alone, briefly popping back into Germany before continuing through the Dutch countryside into Belgium. She traveled mostly at night, covering up to forty miles at a stretch. By day, she laid low. The female wolf, whom biologists named Naya, settled on a military base near Leopoldsburg. She had covered more than three hundred miles in ten days. We know all this because her movements were followed closely by Hugh Jansman from his base in Wageningen. En route, she killed a handful of sheep but mostly stayed out of trouble. When she ended her odyssey, Naya became the first wolf to take up residence in Belgium for more than a century.

Naya did not have to wait long for company. In August, a male wolf (later named August) joined her after making the same trek south. The two kept to themselves, and by the following spring, August was seen taking food back to Naya, a sign she had given birth to pups. But in June, Naya disappeared. The last time she was seen alive was on a wildlife camera operated by the Flemish Nature Agency. Environmentalists suspect hunters killed her, a claim that hunting organizations vigorously denied. A reward for information did not produce any conclusive evidence. Her death was widely condemned. The Belgian office of the Worldwide Fund for Nature issued a statement declaring, "The death of the wolf and her pups is a shame for Belgium."

August lived a lonely few months after Naya was killed. He stayed out of trouble and mostly kept a low profile. This changed on Christmas Day 2019, when a kangaroo went missing in Balen in northeastern Belgium. It is not often that a wolf gets to eat a kangaroo. Wolves don't live in the Southern Hemisphere, and there aren't many kangaroos in the north. This one lived as a pet in someone's back garden. Biologists are confident August was the culprit. Jan Loos, the director of Welkom Wolf, found tracks matching August's at the site. August was known to be in the area, keeping out of sight by day and making opportunistic strikes at night. He was doing what wolves do, taking advantage of easy prey before returning to the shadows.

On New Year's Eve, the *Brussels Times* reported that a new wolf, Noëlla, had been spotted on wildlife cameras in Flanders near where August lived. Noëlla and August were seen together frequently in the Bosland Nature Reserve over the next few weeks. The timing was perfect. Breeding season started in February. Before long, cameras showed Noëlla was pregnant. She gave birth to four pups in May 2020. Eddy Ulenaers at the Flemish Agency for Nature and Woodlands said his heart jumped out of his chest when he saw the pictures. Local authorities, wanting to avoid what happened to Naya, introduced a hunting ban. Wildlife rangers protecting the pack were offered support from the Belgian military. Times in Belgium had changed.

August became a minor celebrity. Schoolchildren followed his movements closely. It was not just the fact he was the first wolf to breed in Belgium for a century. He was also a recognizable individual who had faced down a series of trials and challenges. Opsomer stressed to me again how Belgians identify with figures who do not respect borders, admiring those who make their own way in a strange land.

Wolves like Naya, August, and Noëlla are easy to identify with. Their lives contain tragedy and loss. They are filled with doses of intrigue. Sometimes, there are elements of comedy. It is not hard to see something noble in what they achieve. Naya opened a fresh chapter for wildlife in Belgium. She cut a path and took risks. She rode her luck for as long as she could until her luck ran out. Wolves made it back to Belgium after a century because of individuals like Naya. These are tales not of insensitive machines, as Descartes might have believed, but of venturous souls writing new stories for their kind. If you are moved by this way of thinking about animals, you might view such stories as heroic. Veteran Yellowstone wolf-watcher Rick McIntyre would no doubt prefer the term *Shakespearean*. And who, in the end, would disagree with him?

PRAIRIE PURITANS

1 A BISON IN A BOXCAR

It was late spring, and I was watching two hundred bison graze by the side of the road on the Blackfeet Reservation near Glacier National Park. To the east, the gentle undulations of Montana's high line wrinkled five hundred miles of sparsely populated prairie. To the west, the elevated crags of the Rocky Mountain Front announced the western boundary of the grassy tableaux. Today the bison were moving gently toward the high country, one mouthful of grasses at a time. Snow-clad peaks, wind-ruffled vegetation, and bison pulling on abundant grasses. Everything felt as it should.

The sight before me would move anyone with hope for recovering wildlife and a gentle ache for the weather-worn tilt of the Great Plains. But beneath the photogenic veneer dwelt a puzzle that burdens a wide swathe of wildlife returns. Can you be certain about what counts as a species recovery? With a cold breeze on my neck and no sign of human inhabitation for many miles, the bison stretching across the prairie in front of me certainly looked like a species doing well again. But with bison, I was about to learn, appearances can be deceiving. There is much more to their recovery than one idyllic western panorama can reveal.

* * *

The sight of bison on tribal lands calls to mind two historic tragedies. One shattered the lives of Plains Indians. The other nearly destroyed the American bison. To the white settler, the Indians on the Plains were an impediment to their westward-facing dreams. Europeans disembarked from covered wagons and railroads onto prairie they thought begged for the plow. Advertisements back East told them this land was their future. They regarded the native people who lived there as obstacles in the way.

The sixty million bison the settlers found seemed inexhaustible. Herds stretched dozens of miles from one end to the other. Paddle steamers plying the Missouri stopped for half a day while tens of thousands of bison crossed in front of them. When they ran, their hooves created a thunder that echoed through the ground for miles. Families resting in teepees and covered wagons could hear the rumble long before the first animals appeared over the bluffs.

Despite their numbers, the bison were no match for the settler's rifles. The prairie's most perfectly adapted tenant was an easy source of meat and hides for the aggressive newcomers. The European appetite for commerce was so voracious white hunters often took only a tongue or a hide for market, leaving the rest to rot where it fell. Wolf numbers ballooned as they feasted on piles of decaying meat. The settlers sent skins back to the East Coast, where they were turned into clothing and leather. The tough hides were fashioned into drive belts for factory machinery. When there were more carcasses on the prairie than bison, the settlers collected the bones and ground them into fertilizer. Aristocrats in Europe drank tea out of porcelain cups fired from bison bone china. They sweetened it with sugar filtered through bison bone char. The white man treated the bison as completely expendable. Railroad owners mounted guns on trains for tourists to take potshots as their carriages rattled by. Even sixty million bison could not withstand this level of callousness.

When the bison suffered, so did the Plains tribes. Blackfeet elders say the tribe's name comes from the color of the hooves of the animal they call the *iinnii* or buffalo. It wasn't long before those who wanted to rid the plains of Indians saw the destruction of the buffalo as the perfect means to do so. In the 1870s, U.S. General Philip Sheridan suggested wiping out the buffalo was "the only way to bring a lasting peace and allow civilization to survive." The general lobbied for medals for successful hunters with a dead buffalo on one side and a discouraged Indian on the other. The U.S. government was happy to see buffalo and Plains Indians decline together. President Grant's Secretary of the Interior insisted the civilization of the Indian would be impossible while buffalo remained alive. Grant vetoed a bill passed by both houses of Congress to protect the dwindling herds.

Bison numbers plummeted to less than a thousand in four short decades of bloodletting. At the peak of the butchery, commercial hunters shot five thousand animals a day. A railroad engineer reported it was possible to travel one hundred miles along the Santa Fe's tracks stepping from one carcass to another along a causeway of death. Decimated by disease and killed off in skirmishes, the remaining Indians were corralled onto reservations. By century's end, the Plains tribes hung on in a tiny fraction of the territory they used to occupy. It was one of the most brutal destructions of people and wildlife in human history.

The bison I saw grazing by the highway that day conjured up both of these tragedies. Yet they also carried with them a vision of hope. Reservations across the Great Plains support a growing number of *innii*. The Blackfeet are playing a vital role in this rebirth. Outside Native reservations, bison numbers are also swelling. It's a huge conservation success, but it comes with a twist with far-reaching implications for species recovery everywhere.

It turns out the bison grazing by Highway 2 are not the Blackfeet's most cherished animals. The ones I saw are the tribe's commercial herd, managed as livestock and shot through with cattle genes. For most conservationists, the real excitement lies somewhere not visible from the highway where a smaller herd of bison also graze their way across unbroken prairie. These animals have a far greater promise of renewal riding on their shoulders. With Plains bison, it's not enough simply to look like a bison. You also need the genes.

* * *

"I have to admit, I'm a coffee snob."

I was sitting across a wooden picnic table from Roy Bigcrane on the Flathead Indian Reservation. It was a bluebird morning in mid-July. A mile away, the towering peaks of the Mission Mountains held remnants of the winter's snow, creating a perfect backdrop for our conversation.

Bigcrane asked if I could pause my notetaking for a moment while he jogged back to his car for a refill. "Two cups of dark roast a day keep me going," Bigcrane announced on his return. "Caffeinated!" He took a sip and let out a contented breath, grinning a toothy grin beneath blue reflective glasses and a backward-facing baseball cap. A long ponytail streaked with gray trailed down his back.

I had come to the Flathead Reservation because Bigcrane has a personal connection to the story of bison recovery. "Their blood is flowing through my veins," he said. His great-great-grandfather was Atatiće (Peregrine Falcon Robe), the Pend d'Oreille Indian reputed to be the first person on the continent to try to save bison. Bigcrane works as a media assistant at nearby Salish-Kootenai College and starred in a documentary about his ancestor. When I sent him an email about the film, he graciously agreed to help me understand the story.

Our rendezvous spot was fortuitous. The Mission Valley is not only where Peregrine Falcon Robe lived; it is also home to the Bison Range (formerly the National Bison Range). This federal facility, established with the help of wealthy conservationists from the East Coast, was another key piece in the story of bison restoration. I was in luck. One valley. Two great ties to the story of bison.

Bigcrane talks thoughtfully with a resonant, oaky voice. But he didn't mince his words when I asked about the nearby Bison Range. "You mean the land they stole from us?" he asked. The Bison Range, as Bigcrane sees it, has always been a finger poked in the tribes' collective eye. When first proposed in the early 1900s, the Salish, Kootenai, and Pend d'Oreille tribes had already spent fifty years condemned to a reservation. The government delivered another blow by declaring parts of the reservation surplus and available for sale to white settlers. The proposal for a Bison Range meant the tribe was now set to lose eighteen thousand acres of what little they had left.

"It's high up there on the list of grievances," Bigcrane told me. "We already gave up a lot of land in western Montana to have this reserved for us." he says. "It's as if you have your nice house and your land and all of a sudden someone says, 'You just have to stay in the bathroom. You can't go into the living room or bedroom. You can't go onto your porch.' That's what it was like for us. The land was stolen." The Bison Range was the final straw for a people who look at land not as a commodity but as source of spiritual sustenance.

Exactly why the government wanted to put a Bison Range on the Flathead Indian Reservation is a bit of a puzzle. The Mission Valley is not prime bison habitat. It was never home to the giant herds that blanketed the prairie a hundred miles to the east. It's a small tabletop of terrain encircled by several mountain ranges. Half of the table is taken up by a giant freshwater lake. The answer to the puzzle about why the range is here turns out to be intimately connected to the deeds of Bigcrane's ancestor.

* * *

The Pend d'Oreille Indians were constantly on the move, crisscrossing the region in a well-rehearsed annual rhythm. They knew the best places to catch fish, hunt bison, and gather plants. They also knew exactly the seasons to be there. The Pend d'Oreille continued to move to this rhythm as best they could even after the 1855 Hellgate Treaty forced them onto the confines of the Flathead Reservation.

Bigcrane's great-great-grandfather, Atatiće, was part of a Pend d'Oreille hunting party that traveled east from the Flathead over the mountains into Blackfeet territory in the 1860s. When hunting was over, Atatiće asked the elders if they could bring some live bison home to keep in the Flathead. The elders weren't sure. They knew the bison herds were diminishing. It was getting harder and harder to hunt them. But the trip east of the mountains was part of the annual routine. The Plains offered a place to relax, an opportunity to gather foods, and a chance to meet—or make war on—other tribes. The elders did not want to give this up. To Atatiće's disappointment, they returned home without any animals.

Atatiće's son, Łatatí (Little Falcon Robe), always wanted to realize his father's dream. On a trip to the same hunting grounds a decade later, Łatatí brought home half a dozen buffalo calves. Tribal members still take pride in the skill it took him to guide the young bison through the steep mountains while dodging bears, wolves, and cougars. The tribe welcomed the new additions to the reservation. By giving them refuge in the Flathead Valley, they were paying the bison back for all they had done. Little Falcon Robe looked after them well, and the herd began to grow.

In the early 1880s, the herd, now numbering eighteen, was bought by Michel Pablo and Charles Allard. Pablo was half Blackfeet-Piegan and half Mexican and one of the most successful ranchers in the region. Allard was his neighbor. Both of them had a fondness for wild bison. But their purchase was not just nostalgic. It was an investment. The ranchers could see the writing on the wall. The animals on the plains were almost gone.

The Pablo-Allard herd lived fence-free along the gentle curves of the Flathead River for more than two decades. While bison continued their downward spiral across the continent, the descendants of Łatatí's bison grew until they numbered close to seven hundred, the largest herd in the country. When bison in Yellowstone sank to less than two dozen at the turn of the century, the U.S. government purchased eighteen from the Flathead to provide Yellowstone a lifeline. An animal from the Pablo-Allard herd was the most authentic Plains bison you could find anymore.

The turn of the century marked the beginning of the end for the Pablo-Allard herd. The Flathead Valley was coming under increasing development pressure from white settlers. The U.S. government allotted some of the reservation's acreage to Indian families and put the rest up for sale. The Hellgate Treaty was blatantly ignored. The herd's future looked precarious.

Allard's relatives had already sold their share of the herd after Allard's death in 1896. Some of them ended up on a ranch owned by Charles Conrad north of the reservation near Kalispell. Pablo now offered the U.S. government the rest of his animals. They ignored him. He eventually found a willing buyer in the Canadian government, which, at the time, appeared to take bison conservation more seriously than its southern neighbor.

With his buyer secured, Pablo hired the most experienced cowboys in the region for a roundup. It wasn't easy to corral the feisty animals for the train ride to Canada. One group was driven to the base of a cliff, where the cowboys thought they had them cornered. The bison climbed the cliff like mountain goats and made their escape. The first bison loaded onto a boxcar simply walked to the far end, broke through the timbers, and jumped out the other side. The roundup spilled over to several summers. Visitors came from across the region to watch, including the artist Charles M. Russell, who painted a journalist from Butte losing his pants by jumping into a tree while escaping a stampede.

After six summers, the last of the herd was shipped off the reservation. Bigcrane's ancestors grieved. Some of the bison ended up at Elk Island National Park in Alberta. There they bided their time. The descendants of the bison brought to the Flathead by Bigcrane's ancestor were special. They were genetically unique and had lived wild for several decades. In Canada, they were given good habitat and the protection they needed to multiply. Łatatí's bison, temporarily in exile north of the border, were to become the seeds of a remarkable comeback.

* * *

Plains bison (*Bison bison bison*) are one of two subspecies of bison in North America. They are what most people bring to mind when they visualize a bison. The larger woods bison (*Bison bison athabascae*) has a squarer hump and less of the thick cape of hair around its shoulders. It lives farther north in Canada. A third bison, *Bison bonasus*, lives in Europe.

The word *bison* comes from a Greek word meaning "original wild ox." With short horns, muscled shoulders, and steely expressions, they look like an animal that could pull a plough. There is a line of modern farm equipment called The Bison. But the imagery deceives. The bison's wild streak means they never proved very suitable for the yoke.

Plains bison evolved from several earlier forms of North American bison. *Bison priscus*, the steppe bison, migrated across the Bering land bridge 195,000 years ago when retreating ice created an exposed plateau between Asia and North America. *Bison latifrons*, the long-horned bison, was a descendant of the steppe bison and one of the largest grazers ever known. *Latifrons* was twice the size of a modern bison, standing eight feet at the shoulder and sporting horns seven feet across. There were a couple more evolutionary steps before *Bison bison* appeared. With each step, the bison became smaller and nimbler in response to hunting by early humans. Bison had to adapt if they were to live alongside a new predator with opposable thumbs, a large brain, and a growing arsenal of weapons. Even in its shrunken evolutionary form, *Bison bison* still outweighs another surviving giant from the Pleistocene, the musk ox. They remain the largest land mammal in North America.

Bison live in mixed-age herds. Like rowdy teenagers, the larger bulls tend to go off on their own or in small male-only groups until hormones draw them back to the herd. The females become sexually mature between the age of two and four. They have an estrous cycle of three weeks, and most females become pregnant during a short window in late July or early August. They give birth, generally to a single calf, nine months later.

Young bison are precocious. The calves can walk within minutes, and they are ready to supplement their mother's milk with grasses within a few days. Rapid development is a big advantage on the unforgiving plains. The youngsters lack the shoulder hump of the adults and look more like young cattle. Their fur is lighter than their parents and has an orange tint, earning them the nickname "red dogs."

While the males are generally disinterested in the fate of their young, the females are attentive parents. The older ones are the matriarchs of each bison group. If the herd has to cross a river, a female will swim upstream of her calf so the youngster has a protective eddy. When fleeing danger, the calves are put on the inside of the thundering swarm of bison to be farther from harm's way. If threatened by predators, the adults form a protective circle around the smaller animals with their horns facing outward. Their short back legs and slender rump mean they can pivot rapidly on their front legs. They are agile fighters. The mother's horns are sharper than the males' and have a hook prefect for launching a threatening wolf or mountain lion twenty feet into the air.

Bison also know how to run. They have large lungs and tracheas for efficient oxygen circulation. They are remarkably athletic for an animal weighing close to a ton. They can hit thirty-five miles an hour and jump over obstacles six feet high. There is only one wolf pack in Yellowstone that regularly takes down bison. For the rest, it is just not worth the risk. All in all, a bison's odds on the right landscape are good. A herd can grow by 17 percent a year. One upshot of their toughness, their athleticism, and their evolutionary refinements is this: bison are, and always have been, a species primed for recovery.

* * *

Atatiće and Łatatí may have been the first, but they weren't the only people determined to save the bison. Shortly before Pablo sold his Flathead herd to Canada, wealthy conservationists on the East Coast started to become serious about protecting the vanishing beast. The country had a conservationist president in Theodore Roosevelt, and environmentalism was taking off. Harriet Hemenway and Minna Hall had formed the Massachusetts Audubon Society in 1896 to protect wading birds. New York's Adirondack Park had been protected. On the west coast, the Sierra Club was advocating hard for federal protection of Yosemite under the charismatic leadership of John Muir. A growing chorus of environmentalists was fighting to slow the relentless exploitation of the nation's forests and wildlife.

Roosevelt saw conservation through a hunter's eyes. In the 1880s, while still a New York legislator, he had traveled to the Dakotas to shoot bison. On his return, he teamed up with George Bird Grinnell and founded the Boone and Crockett Club to promote the management and protection of wildlife for hunting. Though he liked to kill animals—something his friend John Muir later badgered him about—Roosevelt recognized that hunting couldn't happen without healthy populations of wildlife.

In 1905, he joined William Hornaday and Ernest Harold Baynes to form the American Bison Society.

Hornaday, like Roosevelt, was an explorer, an outdoorsman, and a naturalist. After abandoning his study of taxidermy in college, Hornaday traveled to India, Borneo, and South America, where he observed, shot, and stuffed almost every animal he saw. He came home telling swash-buckling tales about jaguars, elephants, and crocodiles. It was after he took a job as a taxidermist at the National Museum in 1882 that he started to be concerned about America's dwindling bison. He persuaded his employer to let him mount an expedition to Montana Territory in the mid-1880s to collect a few remaining specimens for the national archives. With a flair for melodrama, he called the trip "The Last Buffalo Hunt." Bison were by now so rare it took seven weeks scouting the prairie before the expedition even glimpsed a live animal. Over the next few weeks, they killed twenty bison, stuffing six of them for the nation's collection. These icons of the plains were already so beleaguered Hornaday found old bullet fragments in every animal he stuffed. He fully expected the species to go extinct.

In addition to the animals he taxidermied, Hornaday also brought back live bison, which he put in a paddock behind the Smithsonian building on the Washington Mall. These animals, some of which originated from the Pablo-Allard herd, were moved to the newly opened National Zoological Park in 1889. Over the next few years, Hornaday conducted a national bison census by visiting ranches, examining county records, and talking to cattle and bison experts. His best estimate was that there were only 541 bison left in the United States. A few hundred more held on in Canada. After becoming founding director of the Bronx Zoo in 1896, he brought forty bison to the new zoo in 1899 as the seeds of a conservation herd. Some of these were Pablo-Allard bison from his Montana travels.

Hornaday's time at the zoo marked a new era for conservation, but his tenure was marred by his association with prominent racists and eugenicists. At one point, he put a villager from the Congo, Ota Benga, on display in the primate house. When he was criticized for his racism, Hornaday claimed he was simply being scientific. Hornaday's attitude matched that of many leading environmentalists at the time. Nature was pure in a way only the white man had the capacity to appreciate. Conservation was the privilege of Hornaday's race to pursue. The taxidermist never apologized for his racism, and the environmental movement around him was unwilling (or unable) to see it. In the meantime, the American Bison Society had no qualms about centering itself at the Bronx Zoo and anointing Hornaday as its dean.

With a new set of advocates across the country and stricter protection enforced in Yellowstone Park, bison herds began to grow. Thanks to a procession of executive orders from Roosevelt's pen, America had a growing collection of national wildlife refuges. Some of the refuges wanted bison. The Bison Society used animals from the Bronx Zoo to conduct the country's first wildlife restoration in 1907, sending fifteen bison to Oklahoma's Wichita Mountains Wildlife Refuge. The society followed this up by sending fourteen to Wind Cave National Park in South Dakota. At the same time, it started scouting locations for a National Bison Range to serve as a western hub for its conservation efforts. Congress appropriated the money and hired a wildlife expert from the University of Montana named Morton J. Elrod to find a suitable location. Elrod chose a site encircling Red Sleep Mountain near Moiese, Montana.

The selection of the site was either a cruel irony or a genuine nod of gratitude toward the Pend d'Oreille tribe. It was a stone's throw from where Little Falcon Robe had lived on the Flathead Reservation. To stock the new national herd, the Bison Society bought thirty-four animals from Lettie Conrad, widow of the rancher who acquired part of the Pablo-Allard herd on Allard's death. The National Bison Range, like Elk Island National Park in Canada, was stocked almost entirely with animals descended from the ones Little Falcon Robe had led across the mountains. The first bison arrived on the new refuge in 1909. Eleven females gave birth the following spring.

* * *

Although the Salish and Pend d'Oreille had been stung by the land grab that created the Bison Range, they were happy to see the animal back.

"The notion of bringing bison back and bringing them back to the reservation was seen as very positive," said Germaine White in the Atatiće documentary in which Bigcrane had appeared, "until they put a fence around it." The tribe were further shocked to discover regulations prohibiting Indians from working at the refuge. The federal government clearly did not see this as a matter of returning bison to their original caregivers. It was a recovery project done by white men for white tourists.

The word the Salish coined for the Bison Range pulled no punches. *Nłox̣ʷenč* means "fenced-in place." The fences encircling the bison range felt as though they were there as much to keep the Indians out as to keep the bison in. The fact the tribe saved the ancestors of the bison that stocked the range was an additional sting. The pain was still sharp a century later. The only just solution, Bigcrane told me, would be a full return of the land to tribal ownership. "Full ownership is what I think would be fair," he said, looking me straight in the eye. "It's our land."

As the injustices piled high and the politics raged on, the bison on the flanks of Red Sleep Mountain focused on doing what they do best. They rapidly multiplied their numbers. By the late 1920s, there were seven hundred bison at the National Bison Range. Excess bison could now be shipped to other refuges to start conservation herds.

The Bison Range wasn't the only place where the numbers were taking off. The Elk Island herd did so well the park started sharing its bison in the 1920s. More than 250 miles to the southeast of the Bison Range, the Yellowstone herd quickly recovered to 950 animals. When overpopulation in the Lamar Valley became a problem, the park started exporting bison to zoos and refuges. These three parks became the three anchors of bison recovery. It all happened very rapidly. By 1935, the American Bison Society was confident enough in the bison's future to vote itself out of existence.

The success continued. Today, 120 years after their low point, *Bison bison* number over half a million. In 2016, President Obama signed an executive order making the bison the national mammal of the United States. Eighty percent of the bison in conservation herds in North America are descendants of Elk Island stock. This means four out of five wild bison contain the genetics of the animals brought to the Flathead Reservation by Roy Bigcrane's great-grandfather.

Seven months after Bigcrane and I spoke in the shadow of the Mission Mountains and 113 years after the original theft of the land, the tribe finally received a fragment of justice. The federal government signed an agreement to begin the process of transferring management of the bison range back to the tribe. Title of the land would move from the U.S. Fish and Wildlife Service to the Bureau of Indian Affairs. The transfer happened through the back door, riding on the coat tails of a congressional bill settling an unrelated dispute about water rights. Bigcrane could exhale now. A century-long wait was over.

There is undoubtedly much to celebrate. But while this sounds like an unqualified wildlife success with a mote of social justice thrown in, a curious fact about bison means the picture is considerably more complicated than it first appears. Many of the bison alive today are genetically compromised. A bison can be riddled with cattle genes and still look exactly like a bison. Genetically pure bison, according to many biologists, are the real deal. Bison with cattle genes are a fraud. As a result, the conservation status of bison is more tenuous than it first appears. To have a sense of just how tenuous, you need to step away from the bison you find in parks and refuges and go visit a bison ranch.

2 A SAUSAGE ON A STICK

Words don't come easily when you are looking at the side of a barn with eight-foot planks ripped apart and left as kindling at your feet. The ranch manager standing next to me on a crisp October morning wasn't a big talker. But he knew how to get to the point.

"I think we got a problem bear."

It was the bear's second visit in two nights. The previous night, it had broken into a different outbuilding and raided a freezer, strewing packets of frozen elk across the gravel outside. Montana Fish Wildlife and Parks had come by and set up a ten-foot-wide electric doormat to discourage the grizzly from trying the freezer again. Last night, the bear looked elsewhere. The wildlife agency would probably have to trap it and move it to the high country, hoping to persuade the bear to make its living farther away from people. We were staring at an occupational hazard of ranching in the Rocky Mountain West. The break-in did, however, provide a way into a topic I had wanted to ask about. When you ranch bison, do you worry much about predators?

Ryan Nemmers, the forty-year-old ranch manager, stood impassively with wrap-around sunglasses and hands buried deep in his jacket pockets. He shook his head. "Elk are the biggest pain in my ass," he said. "They push down fences or break holes in them. I hate elk on the ranch." As the evidence in front of us suggested, bears were more likely to damage farm property than harass Nemmers's herd of 215 bison.

I was surprised the bears and wolves occupied so little of Nemmers's time. The Chamberlain Mountain wolf pack lived just over the hill from the ranch. What about when the calves are born in early summer?

"Nope. Bison are one of the most protective animals with babies. In the first couple of weeks, the calves stay less than ten feet from their mothers. Ninety percent of the time, they are right next to or underneath their moms." If trouble did come, the bison were pretty good at taking care of themselves. "The adults are more agile than people think," Nemmers said. "Most of the time, it's not worth the predator's effort." This was not the case, he told me, for a neighboring cattle rancher a couple of miles up the creek who had lost a few calves to wolves.

The bison's hardiness is part of why Nemmers loves working with them. In summer, he rotates them through the ranch's four different pastures, where they spend three months grazing on their own. In winter, he feeds them giant bales of hay he cuts from his fields each August. Two years ago, it took him four hours with the tractor to plough the eight-foot snow drifts blocking his way to the barn. Aside from the hay in winter, the bison more or less took care of themselves.

"They mostly do their own thing" he said. They stay healthier than cattle, the winters don't faze them, and he never has to help them give birth.

"In fact, I've yet to see one born," he said. "They go off on their own somewhere. They'd probably maul you up if you did try to interfere." Bison on a ranch, in other words, are much like bison in the wild. They are rugged, well-adapted animals, capable of looking after themselves. "Cattle are more work!" said Nemmers. And he knows. He has lived in this area working on ranches his whole life.

* * *

When bison numbers plummeted at the end of the nineteenth century, more was lost than the jaw-dropping numbers of the wild herds. The remaining animals also started to lose their genetic identity. Cattle ranchers noticed the occasional cross occurring between bison and cows.

This perked their interest. There was plenty to admire in a wild bison. They were muscular, independent, and hardy. They reproduced reliably and flourished on grasslands too dry for most cattle in summer. They also knew how to survive the bitter winters of the Great Plains. Ranchers soon worked out that breeding bison traits into cattle could provide an advantage in a tough business. As wild bison declined, entrepreneurial cattlemen in Kansas, Texas, and the Dakotas got their hands on some of the last remaining animals to experiment with cross-breeding.

Bison and cattle form an odd couple. They are not the same species. They are not even the same genus; cattle are in *Bos*, and bison are in *Bison*. Despite the difference in lineage, they are able to interbreed successfully. When they did, ranchers liked what they saw. Ernest Seton, a turn-of-the-century naturalist, wrote at the time of the first crosses,

> The hybrid animal is to be a great improvement on both of its progenitors, as it is more docile and a better milker than the Buffalo, but retains its hardihood, while the robe is finer, darker and more even, and the general shape of the animal is improved by the reduction of the hump and increased proportion of the hind-quarters.

The challenge was to create hybrid animals that could reproduce. For a long time, only female crosses could breed. It took decades of experimentation before a rancher in Montana created a hybrid bull that was reliably fertile. Farmers no longer had to keep unruly wild bison on hand to fertilize the females. Beefalo became an official strain of cattle marked by a genome that was five-eighths cow and three-eighths bison.

Although the 5:3 proportion was necessary to qualify as a beefalo, the number of animals with odd mixes of cattle and bison genes steadily increased as more and more experiments in cross-breeding took place. Much of the genetic dilution took place under the radar. Bison can contain a lot of cattle genes before they stop looking like bison. Knowing whether an animal is genetically pure by sight is more or less impossible. Genome sequencing was still decades in the future. By the time the technology arrived, cattle genes had made their way into the vast majority of bison herds.

The mixing of bison and cattle genes does not matter much from the point of view of a rancher. You have a rugged animal that meets your market needs. But cattle genes matter a lot from the point of view of bison conservation. If you are trying to save a vanishing species, genetic integrity is often thought to be the heart of it. Michael Soulé, a founder of the discipline of conservation biology, insisted that one of the most important beliefs in conservation is that the results of evolution are good. "Evolution is the machine, and life is its product," he claimed. This squared nicely with more than a century of conservation thinking. The genetics of a species count. "Species conservation is more than skin deep," says James Derr, a professor in the department of veterinary pathology at Texas A&M University. "It is more than how they look, it is how they are—that's the genome." Hybrid species—and especially species hybridized to satisfy human needs—are valued less than species with their original genetic legacy intact. The reasoning does have its appeal. There are more than two thousand centuries of hard-won, North American experience baked into the genome of a wild bison. An intact connection to *Bison priscus* and *Bison latifrons* means something.

When we drove over to look at his herd, Nemmers pointed out the four huge bulls who sired most of the forty calves that year. The biggest, known affectionately as Semen, chewed hay as he lay on the ground looking straight at us. His massive shoulder hump blotted out the animals behind him with a wall of muscle. "He's really friendly," said Nemmers. "He was bottle-raised by the previous manager."

I asked Nemmers if he knew the genetic make-up of his bison. "To be honest, no," he said. The owners of the property bought a couple of dozen bison twenty-five years ago before genetic testing was widely available. They reproduced so successfully the ranch never needed to shop for more. The problem constantly on Nemmers's mind was how to stop the herd growing too big for the available land. "Forty calves this year means forty bison I've got to get rid of," he told me with a frown. The question of genetic purity never comes up. By avoiding corrals and annual inoculations, Nemmers runs the bison as naturally as he can. DNA testing is expensive, and there isn't much to gain by doing it.

As we drove around the edge of the herd in the four-wheeler, Nemmers pointed out the big males, the breeding females, and the season's young. One calf, born unusually late, still sported the orange coat of the newborns. Most of the bison looked like they would fit as well in Yellowstone National Park as they did right here in western Montana's ranch country, a stone's throw from the Blackfoot River. Even though they looked authentic, to suggest these animals might be interchangeable with Yellowstone bison would be, for many people, to commit wildlife heresy. Conservation's focus on genetic purity would frown upon it. It would be like saying the difference between your neighbor's kitty and a Himalayan snow leopard didn't matter.

As bison recovery continued and genome sequencing became more widespread, the question of genetic heritage became increasingly important to conservationists. You might have something that looks enough like a bison to impress a visitor at a state park. Yet beneath the matted fur, the dark eyes, and the towering shoulder hump, it was not a real bison. Of the half million bison now grazing across the United States, most of them are hybrids. Many of these are destined for slaughterhouses, where they will be turned into steaks and burgers. Only thirty thousand exist for the purpose of conservation. And of these thirty thousand, just a fraction of them are thought to be free of cattle genes. This has the potential to upend the conservation success story. True bison, whatever that means, remain perilously few and far between.

* * *

"I bet Yellowstone doesn't have genetically pure animals. Nobody knows. They have never tested them."

I was sitting in a conference room at American Prairie's headquarters in Bozeman, Montana. Across the table from me was Pete Geddes, vice president and chief external relations officer of the upstart conservation outfit. The father of three grown boys, Geddes previously worked at the Property and Environment Research Center, an organization that calls itself "the home of free market environmentalism." Some of the same free-market attitude percolates through American Prairie. Geddes likes to approach his environmental work using market mechanisms and philanthropy rather than government edict and law enforcement.

At first, the former outdoor leadership instructor came across as really low-key. He kicked back in his chair and delivered his words in a casual riff. We started by talking about the hockey game he had to be at in an hour. It was late on a Friday afternoon, and I wondered if Geddes already had one foot in the weekend. But as soon as the topic turned to prairie restoration, Geddes turned laser sharp. He leaned forward in his chair, and the words flew like arrows. In a part of the state where cattle is king, what American Prairie plans to do with bison needs very careful articulation. The long-term vision of a 3.5-million-acre refuge, managed in a public-private partnership, requires patience, political savvy, and the careful building of trust.

American Prairie is slowly cobbling together a patchwork of land and grazing leases around the 1.5 million acres of the Charles M. Russell National Wildlife Refuge and the Upper Missouri River Breaks National Monument. These two publicly owned sanctuaries will anchor a continuous spread of protected land more than 50 percent bigger than Yellowstone. The organization promises "the largest nature reserve in the contiguous United States, a refuge for people and wildlife preserved forever as part of America's heritage." Done right, it would turn part of central Montana into a homespun Serengeti with huge herds of ungulates dodging a suite of predators on a landscape scale. Bison would be the lynchpin holding it all together.

Geddes's claim about the genetics of Yellowstone bison is at odds with what you normally hear in these parts. Bozeman sits within a hundred miles of two of Yellowstone Park's entrances. The park's bison are widely revered in this college town, which serves as both an outdoor mecca and a landing pad for movie stars. Yellowstone bison are the only bison to have lived wild and free on the same landscape since the slaughter ended a century ago. Their mystique has garnered them their own environmental group, the Buffalo Field Campaign. It's a mystique they thoroughly deserve. A big bull bison, frosted in snow and huffing ice crystals on Yellowstone's Blacktail Plateau, is about as other-worldly as a bison can get.

Even so, the Yellowstone bison have genetic complications woven into their history. They were supplemented at their low point by animals from the Pablo-Allard herd as well as animals from at least one other commercial ranch. For years, the bison were kept semidomesticated at a buffalo ranch in the Lamar Valley. Eventually, the feeding and corralling they underwent there was deemed incompatible with the park's mission, and they were released. The bison in Yellowstone today certainly roam wilder than any other herd in the United States (even though the state of Montana has gone to court to prevent them from roaming *too* far). But nobody knows for sure whether some cattle genes didn't sneak in at some point. As Geddes indicated, nobody has actually tested them. And given that the most significant biological characteristic of Yellowstone bison is they run wild, nobody is likely to do so anytime soon.

American Prairie has stepped in to provide a kind of genetic reassurance to the bison conservation community. "Our goal has always been to have the gold standard for a conservation herd in terms of genetics," Geddes told me. Most American Prairie bison came from Elk Island. A few came from the Wind Cave herd established by the American Bison Society. Every bison is genetically tested by a lab in Missouri before being let onto American Prairie's land. As DNA technologies improve, the animals are retested. Whenever the bison are brought into a corral for an inoculation or a census, blood from a random sample of the herd is sent back to the lab. Animals with cattle genes are killed or given away. The policy is ruthlessly enforced. In 2010, when American Prairie's herd was only two hundred strong, bison managers removed ninety animals because of genetic impurities. The upshot of the intensive management is this: if an animal is grazing on American Prairie's land, you can be pretty confident it contains no detectable cattle genes.

Being sticklers for genetics is an investment American Prairie is making in wild bison's future. For now, these animals are privately owned and treated strictly as livestock according to the dictates of Montana law. But if the legal situation changed, the organization would be ready. Montana Fish, Wildlife, and Parks recently released a report in which it admitted it has an unfulfilled obligation to restore wild bison to the state. The report is vague about how—or when—the state is going to make that happen. But the department went on record saying it must.

"If the people of Montana want a herd of wild bison at some point," Geddes told me diplomatically, "we would be happy to contribute our animals." American Prairie has the golden eggs, should anybody find a suitable place to put them. Until that time comes, the bison managers try to keep the herd pure and slowly build its numbers.

* * *

Danny Kinka's slim and bearded face moves with the energy of someone well suited to the size of the task ahead. Kinka is American Prairie's wildlife restoration manager. He has to think much broader than just bison. Kinka's task is to recreate a prairie heaving with hundreds of species of wildlife. But he is prepared to play the long game. The first step is to create a change in mindset. "We need a reimagining of what this place can look like," he told me. "We did such a swift job of driving all the wildlife into the mountains that there is no cultural memory of this in the white population." The baseline expectation for wildlife has slipped too low. "I view shifting the baseline back upward as a primary component of my job," he said.

American Prairie chose its site carefully with baselines in mind. Eastern Montana is home to one of the last intact temperate grasslands in the world. National Geographic named it as one of its "last wild places" to focus attention on its rarity. Compared to most parts of the Great Plains, a sizable portion of the sod here remains unplowed. Eastern Montana proved a hard place for the immigrants to survive. Even today, it remains a challenging environment in which to make a living.

The founders of American Prairie saw a unique opportunity to stitch together a fully functioning ecosystem replete with as many original species as possible. They envisioned bison, elk, cougars, black-footed ferrets, prairie dogs, bighorn sheep, pronghorn antelope, and swift foxes all returning in numbers. To some extent, it's already happening. The reserve is home to ninety different mammals and more than a thousand plant species. One-hundred-and-fifty-nine types of waterfowl, upland game birds, raptors, and songbirds have been documented on the reserve. Some animals are there in huge numbers. Whitetail and mule deer are close to carrying capacity. Elk numbers are slowly climbing. Swift fox have been reintroduced on the Fort Belknap Reservation to the northeast, and Kinka is hoping they will soon trot their wispy frames south onto American Prairie land. Bighorn sheep are doing well in some parts of the Charlie Russell Refuge, though less well in others.

The reserve is working to grow its existing prairie dog towns. The prairie dogs help maintain the grassland and are essential to the critically endangered black-footed ferret. Staff and volunteers have enhanced the riparian areas by removing small dams and planting native vegetation. They have reintroduced fire to the grasslands to help greater sage grouse, prairie falcons, and burrowing owls. Within a decade, Kinka thinks grizzly bears and wolves will make their way back as they push farther and farther onto the plains from their mountain strongholds.

Bison will play a pivotal role in making it all work. These four-legged ecosystem engineers are a keystone species on the prairie, shaping the landscape as they move across it. Their grazing helps sustain the vegetation. Their hooves aerate the soil. Their dung spreads seeds and adds nitrogen. It is also a magnet for insects. After a fire, bison quickly move into burned areas to enjoy the succulent new grasses, creating a habitat mosaic beneficial for other birds and wildlife. Songbirds use bison fur to line their nests. Pronghorn antelope follow their tracks in deep winter snows, making their own lives a little bit easier.

Geddes told me you can see the bison's impact on overflights of the reserve. The bison make small indentations on the surface when they roll in the dust. These depressions, known as *wallows*, compact the soil and hold rainfall longer than the surrounding prairie. The ephemeral pools serve as a welcome mat for insects and reptiles. The compaction and evaporation they undergo adds diversity to the soil structure.

Bison also shape the land when they die. Their carcasses act like giant furry bags of Miracle-Gro. "You can see from a plane the difference in the plant growth where all the nutrients go into the soil," Geddes said. "Where they die is like a shot of vitamins for the system." A view from five hundred feet shows a land increasingly pockmarked with fertile hotspots.

American Prairie's herd is currently at its limit of just under nine hundred animals. Like the Bison Range and Elk Island, American Prairie is forced to export bison to stay within the carrying capacity of the land. Many of its excess animals go to tribal reservations. In the last few years, it shipped bison to the Nakoda, Lakota, Okáxpa, A'aniiih, and Piikani people. As sovereign nations, the tribes can receive animals from American Prairie without the burdensome regulations imposed on nontribal entities. In a modern-day echo of William Hornaday, American Prairie has also sent animals to the Smithsonian National Museum and the Bronx Zoo. Bison managers conduct a lottery each year for twenty permits to harvest bison. You can only "harvest" livestock, not hunt them, Geddes told me. In this delicate political dance, language matters. The permits are heavily tilted toward locals and tribal members. Only two of them were available in a recent year to people from outside of Montana.

Keeping it local is an attempt at maintaining friendly relations with the neighbors. American Prairie needs all the goodwill it can muster. The fact that it buys out willing ranchers as it pieces together holdings means the organization feels like an existential threat to some of those in the livestock industry. With its Silicon Valley roots and big-donor reputation, the organization is constantly fighting the perception it imposes out-of-state interests on hard-working ranch families.

The bison themselves are caught in the crossfire. America's national mammal is viewed with intense suspicion by cattle ranchers who worry about their potential to transmit disease. Conservative legislators have introduced a slew of bills in Montana's state capitol to make American Prairie's bison operations difficult. Signs on weather-beaten fence posts in the area read "Save the Cowboy, Stop American Prairie." It is too early to say where this hostility will end and what it will mean for the future of wild bison in Montana. But with every new acre purchased and every new bison suckling nourishment from its mother's teat, American Prairie is turning its vision of genetically pure bison into an expanding, cloven-hooved reality.

* * *

In the midst of my immersion in the bison controversy, I stumbled into another story of animal resurgence on Montana's prairie providing conservationists with a measure of hope. If a landscape teeming with bison has social and political hurdles to overcome, one teeming with pronghorn antelope is proving a much easier lift. "Everyone likes them," said Andrew Jakes, a biologist at the time with the National Wildlife Federation. "They are just this magnificent, beautiful creature you would not expect in North America. I have never run into a person who has a tough time with pronghorn."

Antilocapra americana is high on the list for the title of the most elegant creature in North America. Their tricolored coat, inward-curving horns, and finely sculpted bodies call forth images of Africa more than the American west. They don't look like the kind of animal to call some of America's harshest environments their home. Topping out at just over a hundred pounds with thin legs and narrow hooves, they are compact and delicate-looking. Yet they are superbly adapted to the austerity of life on the high plains.

Pronghorn are not true antelope. Their closest living relative is the giraffe. They are the only surviving species in the family *Antilocapridae*. Tens of millions of antelope used to keep company with woolly mammoths, dire wolves, and giant sloths. Thirteen antelope species spent the Pleistocene evading lions and cheetahs across North America. All of them except the pronghorn died out in the megafaunal extinctions.

"Pronghorn," Jakes says, "are built to move." They are the fastest land animal on the continent, capable of upward of forty-five miles an hour. The reason for their great speed, so the story goes, is that they evolved with American cheetahs hot in pursuit. Though the cheetah is now gone, pronghorn have kept their celerity as an embodied memory of the chase. Like bison, pronghorn have enormous trachea and lungs that allow them to maintain their pace for several miles. Over a distance, they can outrun anything on earth. With eyeballs the size of a horse's, they can detect a threat a thousand yards away. When all other defenses against a predator fail, they have an unusual trick up their sleeve: their hair is a good insulator but it pulls out easily. This helps them escape a predator's jaws in a pinch, leaving the would-be hunter with only fur in its teeth and an empty stomach.

Jakes's interest in pronghorn stems in part from the fact he is a leading authority in the field of fence ecology. The discipline is defined in the academic journals as "the empirical investigation of the interactions between fences, wildlife, ecosystems, and societal needs." In simpler terms, it's about keeping people and wildlife happy when their paths cross. Good fence design can go a long way toward soothing wildlife-rancher conflicts. The holy grail in Jakes's line of work is a fence strong enough to keep livestock inside, visible enough for wildlife not to crash into it, and permeable enough for animals like pronghorn to pass through it easily. In the northern regions where Jakes works, it must do all of these things across a range of different snow depths and be resistant to the grass fires that scorch the prairie in summer.

Fence design is an empirically guided form of art. Whitetail deer like to jump. Pronghorn prefer to shimmy. "Pronghorn evolved with the open prairie landscape," Jakes said. "The biggest thing they ever had to jump over was sagebrush. Nine times out of ten, they want to crawl under something." Studies in Alberta show pronghorn crossing fences over a thousand times a year. Jakes has tracked them taking hours to negotiate a single, badly designed fence. He knows it can be done better. "If you raise that bottom wire up high enough, then pronghorn can scoot under it pretty effectively," Jakes assured me. Working with ranchers to implement the right design is essential if pronghorn numbers are to keep growing.

Pronghorn and bison make use of the same territory. Jakes refers to pronghorn as bison's little brother. In winter, when bison use their strength to plow tracks through the snow, pronghorn follow in their wake nibbling on exposed sagebrush. The bison swing their anvil-shaped heads back and forth, grazing up to forty pounds of forage a day in wide swathes. The pronghorn's narrow muzzle allows it to be much more discriminating about its fare. "The pronghorn are much more of a selective feeder," Jakes says. "They have teeth that allow them to select very specific types of forbs and grasses." When spring comes around, many pronghorn migrate to stay up with the emerging vegetation, something biologists refer to as "surfing the green wave." The females migrate while pregnant, keeping pace with the most nutritious grasses to give their newborns the best chance of a good birthweight. Other than Arctic caribou, only mule deer migrate farther in North America.

Although pronghorn did a pretty good job of evading cheetahs, they weren't so adept at staying out the way of white settlers. There were between thirty and sixty million pronghorn on the prairie before settlement. Overhunting reduced this number to thirteen thousand by the early twentieth century. "They are notoriously terrible swimmers," Jakes says, "and their foot-loading index means they are bad at dealing with snow." The decline in bison made movement in winter tougher for pronghorn. If you knew where to lie in wait, you could hunt them pretty easily—so easily, in fact, they became known as "sausage on sticks." The result was a catastrophic plunge in numbers before wildlife agencies and private land owners stepped in with hunting regulations and refuges.

The recovery of antelope since then has been even better than the recovery of bison. The wild population is now close to a million. Jakes is not surprised at their success. "There is a reason they have been around since the Pleistocene," he says. "They can live in a bunch of different types of systems. They almost always twin [at birth]. They do pretty well."

When I ask him how high the pronghorn population could go, Jakes didn't want to commit. There is enough habitat for numbers to be several multiples higher than today. Social tolerance is what will determine how high the number will go. Farmers and ranchers are naturally uneasy about sharing forage with uninvited guests. Learning how to share territory with independently minded mammals takes time and patience. New habits have to be learned. New techniques developed. But in Jakes's mind, the antelope is an ideal candidate for getting people accustomed to cohabitation. Its appetite is much smaller than an elk's, and they aren't so destructive of property. With the right fences, Jakes is confident antelope can thrive and people can enjoy having them around.

"I like trying to find solutions by working with people, solutions that are a win-win for both," Jakes said. When wildlife return, a dedicated cadre of specialists is needed on the ground to ease the pinch points. Jakes relishes being one of these. And there is at least one other consideration that makes his work much easier than if he were working on bison. With pronghorn, the genetic puzzles tend not to come up. All the pronghorn's closest relatives died out in the Pleistocene.

* * *

Despite the fact things are looking up in some sections of the prairie, in others they remain grim. Jim Bailey is growing increasingly antsy about the prospects for *Bison bison*. Bailey is a retired wildlife professor from Colorado State University with an obsession for wild bison. From his home at the foot of the Bridger Range, he runs the Montana Wild Bison Restoration Coalition, where he advocates unceasingly for bison to roam free.

"The species *Bison bison* is secure," he told me. "We have several thousand of them. The threat is domestication." A bison kept behind a fence is no longer subject to the full range of natural forces. Even a bison with an authentic genome, such as the ones from Elk Island, won't stay authentic for long if they are confined like cattle. In Bailey's eyes, fences are as harmful to bison as bullets. "The wild genome," he says, "is being dismantled."

Fences are not the only thing degrading the bison genome. The small size of the herds means their genetic diversity is constantly dwindling. Valuable traits disappear through an evolutionary process known as *genetic drift*. Genes that may be useful don't make it to the next generation through simple bad luck. Not every animal gets to breed. Genetic diversity is always slipping away.

Kinka and his partners at American Prairie are well aware of the problem. They are always on the lookout for sources of genetically diverse bison. But the overall pool is small. With only a handful of genetically pure bison in the United States and many of those protected in Yellowstone, it takes effort to find a bison to spice up your gene pool. The priceless diversity is sprinkled behind fences at various Indian reservations, state parks, and zoos. "There is a concentrated effort," says Kinka, "to try to manage America's remaining bison as a megaherd by periodically moving around an animal or two." A megaherd mentality means thinking of every American bison, from Florida to Alaska, as part of the same breeding population. In practical terms, it means putting bison on trucks and moving them around the country to breed.

Jim Bailey knows the mixing is important, but he thinks bison also need to live wild. The key point for Bailey is to give them the space to roam freely. Without this, they can't live as bison should live. "Mobility is a—perhaps *the*—most basic characteristic of wild bison," Bailey says. Ironically, space became important to bison because they evolved alongside humans. "The evolution of modern bison," Bailey told me, "was a switch from a 'stand and defend' predator response to a strategy based on mobility—in response to the advent of a new predator that threw spears. For natural wild bison, we need a large, diverse range." What Bailey wants, essentially, is for the bison to have a tough time. Winter blizzards, summer heatwaves, difficult births, deathly diseases, predatory cougars, the constant search for food: these should all be part of the bison's experience. It is through facing down challenges that bison maintain their genetic strength.

When you are staring across a fence at a two-thousand-pound bull bison with an imposing brown cape and rakish horns, these finer points about genetics can all seem a bit pedantic. But to a wildlife biologist like Bailey, they matter. Without the relentless whittling of the genome, a bison stops being a bison and becomes something else. In Bailey's mind, bison need to be pure, and they need to live free. Anything else, and you risk losing the animal you are trying to save.

I certainly appreciated Bailey's point about the need for open spaces. But I couldn't help wondering if the bar was being set just a bit too high. When a large animal has been missing from a landscape for a century, it is hard to entirely escape human influence on its return. The ecological conditions have often been missing for decades. The cultural conditions may also have changed. This is especially true when you are talking about places less roomy than eastern Montana.

At the same time, Bailey had neatly highlighted a conundrum swirling around a range of animal returns. If people and fences have to be involved in the recovery of wildlife and if this involvement degrades the animal's authenticity, what does it really mean to restore an iconic animal like a bison? Or to put this another way, when an animal like the bison recovers, can you be sure you have got the right animal back?

3 GENETICALLY AUROCH

The work week had just ended in Europe when I reached Ronald Goderie on Skype at his home in the Netherlands. Though it was still midmorning in the U.S., I watched with a pang of envy as Goderie raised a goblet of Dutch beer to a mouth ringed by a short salt-and-pepper goatee. The slim, middle-aged conservationist took a sip, replaced the glass outside the screen shot, and greeted me with Dutch-accented English and a warm smile. I had been trying for weeks to find time with Goderie to talk about his work on natural grazing. What he and his partners were doing in Europe felt like it contained lessons for any number of large, iconic species coming back from the brink.

Right from the start, Goderie seemed like he was tuned into the same dilemmas I had been mulling over in Montana. "Hunger," he declared, sounding like a European Jim Bailey, "is an essential ecological process. Hunger drives the animals and their behavior." The need for grazing animals to experience hunger was one of the reasons Goderie had spent years developing partnerships in Croatia, Ukraine, and Spain, where large, empty expanses of semiwild landscape call out for the reintroduction of a weighty ungulate. The grazers have to find their own way, their path determined by the tormenting nag of a half-empty stomach.

Goderie runs the Stichting Taurus Foundation, an organization dedicated to restoring natural grazing in protected areas. In most parts of Europe and North America, the landscape can support far more wild herbivores than currently exist. To rectify this, Goderie has spent two decades putting tough domestic breeds like Konik ponies, Bosnian mountain horses, and Highland cattle onto landscapes in need of the pressure of a grazer's teeth. The cattle and ponies do a good job so far as they can, but they lack the jaws and the chewing power some plants require. "Horses graze with their teeth and lips, not with their tongue," Goderie pointed out. "So it is a completely different pattern." The most audacious part of Goderie's work is his effort to fill a gaping ecological hole. The genial Dutchman is fourteen years into a project to cross existing breeds of cattle to bring to life an animal that looks, behaves, and moves through the landscape like the original wild bovine of Eurasia—the majestic but thoroughly extinct auroch.

* * *

The auroch was the ancestor of all domestic cattle. Its Latin name, *Bos premigenius*, translates roughly as "first born cow." Aurochs first appeared in India two million years ago. They ranged across Asia and into North Africa, arriving in Europe around the time the first *Homo sapiens* skull formed between the shoulders of an upright hominid. Early herders domesticated aurochs separately in Asia and Europe during the Neolithic period, creating two main strains of cattle known today as zebu and taurine. The zebu have a fatty hump on their shoulders and a large flap of skin (or dewlap) on their neck. They are well adapted to warm temperatures and drought. The taurines prefer temperate climates and tend to be better milkers.

Large auroch bulls stood six feet at the shoulders and weighed up to three thousand pounds. The horns had the curve of an animal equipped for fighting, broad at the base and arching forward as much as three feet at the tips. Julius Caesar wrote that young hunters coveted auroch horns, which they would "bind at the tips with silver, and use as cups at their most sumptuous entertainments." The horns' weight meant aurochs evolved a muscular neck and shoulders to keep their heads aloft. Battle scars on bones dug from archeological sites suggest males fought viciously to assert dominance over each other. In cave paintings and engravings, aurochs cut an imposing form. But they also look strangely familiar, more like a Spanish fighting bull than an exotic wild animal like a bison.

The auroch grazed European landscapes alongside mammoths, steppe bison, and woolly rhinoceros. They dodged cave bears and lions effectively enough to survive deep into the Holocene. As more and more large animals disappeared due to the warming climate and hunting by early humans, the auroch hung on in rapidly shrinking habitat. It wasn't until 1627 that the last auroch died in the Jaktorów Forest in Poland. Plenty of well-preserved auroch bones have been discovered in forests and peat bogs across Eurasia and North Africa. In 2015, a six-thousand-year-old bone fragment found in Britain became the source of the first successfully sequenced auroch genome. Goderie, if he wants to, can pull up on his computer exactly what an auroch should look like at the genetic level.

* * *

Goderie became interested in cattle after visiting neighborhood farms as a child. "I suspect I picked up a bovine virus of some kind," he deadpanned. He has spent thousands of hours since then closely watching their behavior. His voice fills with enthusiasm when talking about matters as esoteric as the shape of a cow's lips or the musculature in its neck. He is particularly excited by the changes he sees in cattle that live where they can display natural behavior. He calls the process de-domestication. "As you witness the process of de-domestication, you see that cows aren't stupid. They are intelligent," he said. Then he paused. "If you score them on the human intelligence scale, they wouldn't get very far. But we wouldn't score very high on the bovine scale, I guess." He is thrilled by the reemergence he is witnessing of cattle's native intelligence, even though it sometimes gets in the way of his work. "It doesn't make our lives any easier," he complained, "because they become smarter and smarter about not getting caged for roundup."

Goderie studied ecology at university before wildlife genetics became a field. He still feels out of his depth in hypertechnical discussions about DNA. "I'm coauthor of a genetic study on aurochs," he told me at one point, "and I fully understand the summary." On graduation, he started throwing ideas around with a group of conservationists interested in bringing back natural processes to the highly ordered Dutch landscape. The conservationists included Wouter Helmer, who went on to become the cofounder and director of Rewilding Europe. A new generation of Dutch thinkers was realizing you can only have so many dikes, tractors, and chickens packed onto a landscape before the foundations of ecological health start to crumble. "Dutch forests," Goderie complained, "were ecological dead zones." The group wondered how it could fire up Dutch ecology again. One thing its members knew for sure was that their ideas about seasonal flooding, unkempt woodland, and natural grazers were going to rock the existing agricultural boat.

Thinking along the same lines as American Prairie, Goderie decided natural grazing would best be promoted through a private organization. He founded Stichting Taurus to supply herds of self-reliant cattle to landowners interested in restoring degraded landscapes. There were hurdles to overcome from the start. Not unlike the rancher's hostility to bison in Montana, the Dutch farming community was not particularly enthused about semiwild horses and bovines wandering near their cows. But after two decades of careful groundwork and delicate political maneuvering, Goderie's mission has caught the European imagination. "We see things happening that in the past we only dreamt of," Goderie said, wearing the gratified smile of someone starting to see the spoils of a hard-fought battle. "And now sometimes our dreams are being surpassed by the developments that actually take place. We are exporting animals to larger and wilder areas in Europe."

Fifteen years ago, Stichting Taurus came to the conclusion they would never fully restore the Eurpean landscape without the presence of its original bovine. The organization decided to take Europe's wildest domestic breeds and back-cross them to create something as close as possible to the auroch. Biologists mapped the genome of seven promising cattle varieties and charted a pathway back to the original Eurasian bovine from the Pleistocene. If they kept testing the cross-breeds against the DNA blueprint of the auroch, they could track their progress toward the extinct species. Goderie named the future animal the *tauros*.

It wasn't long before Goderie realized there were technical and practical limits to the precision of back-breeding. "It could lead probably to an animal which looked like aurochs, probably also behaved like aurochs, but in the end genetically differed from aurochs," he said. "What's lost is lost." Goderie also knew there had been plenty of genetic diversity within aurochs when they still existed. The recreation of one specific auroch genome would offer only a momentary snapshot.

Goderie slowly shifted his genetic goals. The reasons were political as well as practical. "There is a dark edge on looking back for pure, wild animals," Goderie said. Prior to the Second World War, the Heck brothers in Germany attempted to breed an Aryan übercow modeled on the auroch. One of the Heck brothers was a hunting partner of Hermann Göring, who personally backed the project. The pursuit of genetic purity in animals can, as was the case with William Hornaday, put you in territory dangerously close to eugenics in humans. "Every time you google 'Goderie' and 'Göring,' you get hits," Goderie told me, "and I don't want it. We want to stay as far as possible away from this."

These days, Goderie is infinitely more interested in what the animal does than in how it stacks up against a particular genetic ideal. Like a number of wildlife biologists, he looks at the whole idea of a breed with skepticism. "I think Darwin was actually the guy that invented breeds," he said. "Before that, there were populations with genetic diversity, and that's actually what I think natural species should look like. In the end, nature has to do the selection."

This retreat from genetic purity does not mean Goderie has completely lost interest in aurochs. "To be able to develop biodiversity, it is a good thing when you look as much like aurochs as possible." He is also convinced it is morally appropriate to have an auroch-like herbivore living wild in Europe. "In the end, I think we should say an animal like an auroch is part of the ecological system," he told me. "They belong there."

But he has clearly gotten over the need for genetic precision. I asked how close to the ancient auroch his animals are today. "Over 99 percent," Goderie assured me. But then he pointed out that my own genome was nearly that close to the bonobo monkey's. "It matters," he said, "where the differences lie."

* * *

I wondered if Goderie's shift from resurrecting aurochs to producing an animal whose function mattered more than its genes contained any lessons for what was going on with bison back home. I was familiar with the idea that missing species should be brought back to the landscape to fill an ecological role. I was not as accustomed to the idea the genetics of the species was secondary. I was curious if Goderie's view could shift the bison debate. One particular scientific fact I had blundered across made me suspect it might.

Analysis shows that, in some parts of the world, bovines and bison aren't as separate as you might think. A close look at the European bison's genome (*Bison bonasus*) shows it already contains auroch (*Bos primigenius*) genes. The ancestor of *Bison bonasus*, the steppe bison, hybridized with the auroch 120,000 years ago. The discovery of a missing hybrid ancestor—nicknamed the Higgs-Bison after the Higgs-Boson, a particle posited by theoretical physicists before it was proven to exist—confirmed this cross-breeding had occurred. *Bos*, in other words, already shared some of its genes with *Bison* more than a hundred millennia before enterprising ranchers in North America tried the same thing.

While this crossing was proven to be true only in Eurasia, Goderie mentioned contested evidence that aurochs from Asia had crossed the Bering Land Bridge and bred with bison in North America during the Pleistocene. Beth Shapiro, one of the leading U.S. experts on bison genetics, cast doubt on this theory when I asked her about it. She did concede, however, that steppe bison might have contained bovid genes when they first crossed the land bridge. The evolutionary history of bison challenges even the experts. Like wolves, bison walked back and forth between

North America and Asia more than once during the Pleistocene, making it tough to reconstruct the exact evolutionary history. Even without a definitive answer for North American bison, the knowledge that *Bos* and *Bison* exchanged genes in Eurasia raised an intriguing question. If cattle and bison have mixed and matched before with little evolutionary cost, do today's debates about genetic purity need to be quite so fierce?

* * *

Goderie's thoughts about the importance of genetic purity may have changed, but his commitment to the value of natural grazing has not. "We say now we are aiming at having a really wild bovine again in European nature to do its thing and, by doing its thing, developing and maintaining biodiversity," he told me. He was getting ready to send twenty more tauros to the Lika Plains in the Velebit Mountains of Croatia. The Velebit's dry limestone peaks soar straight out of the azure swells of the Adriatic. It is a diverse ecological landscape, home to chamois, red deer, brown bear, wolf, and lynx. The shifting of the region's economic heartbeat from agriculture to tourism has created an opportunity for wildlife and the emergence of a new, nature-based economy.

The tauros Goderie was sending to Croatia were the second delivery in the last few months. They would join a herd of 160 others already grazing the high-altitude meadows. Goderie understands that for now, the tauros is more likely to be successful in places like the Velebit than at home in the Netherlands. There were too many people walking their dogs through Dutch nature reserves to feel comfortable letting loose a large, wild bovine. "What we need in the Netherlands is aurochs 2.0 with the behavior of a tame dog," he said. Over time, perhaps that could change. But in Croatia, there was more elbow room for the tauros to display its natural feistiness.

The capacity for learning Goderie admired in cattle was already on display in the Lika Plains. The tauros had experienced challenging winters, the all-important hunger, and the threat of predation from the Velebit's wolves. Adult tauros are quite capable of fending for themselves. The young are more vulnerable. In areas without predators, a bovine's natural behavior is to collect the calves into a type of hooved kindergarten with one or two elder cows taking turns to look after the group.

"In the Velebit last summer," Goderie said, "we saw elder cows, but we did not see any calves. This system of a kindergarten seemed not to be there." This puzzled the researchers, who were expecting a crop of forty youngsters. So they asked local farmers to keep their eyes open. They learned the tauros had developed a new strategy. "What the cows seem to be doing is they hide their calf and go there once or twice a day to feed the animal. The calf stays under cover during the best part of the day. This is an adaptation the animals have newly learned in the presence of wolves." As behaviors evolve, Goderie expects to see a sophisticated tauros culture emerge. Natural grazing doesn't only change the landscape. It also changes the culture of the animals that use it.

Beyond Croatia, Goderie is pursuing tauros reintroductions in the Ukraine, the Danube delta, and Spain. He is particularly interested in Spain. The cultural ties to ancient cattle breeds are more alive in Spain than any part of Europe. "Americans think they invented cowboys," Goderie told me. "Well, they didn't, of course. It's a European invention." Cattle culture remains integral to Spanish life. He had met herders still using ancient routes several hundred miles long to move animals between summer and winter pastures, a practice known as *transhumance*.

The various cultural histories meant Goderie had learned to make his sales pitch differently depending on where he was. "In the Netherlands, they are looking at you and saying, 'So you got cows? So what? What's the deal? What's the news?' When you tell them in Croatia, they say, 'Ahhhh, that's interesting. What could be in it for me?' When you tell them in Spain they say, 'Yes, we want it!' Taking ancient cattle breeds to Spain is a bit like returning a missing whisky to Scotland. Goderie hoped to deliver the first shipment of Spanish tauros before the end of the year, confident the auroch-like animal was going to fit right in.

* * *

With Goderie's perspective on genetic purity ringing in my ears, I remembered the animals by the roadside on the Rocky Mountain Front where I first had a taste of bison recovery. I wondered if my reaction at the time had been right. Did the difference between the Blackfeet's commercial herd and its conservation herd matter as much as I had thought? I needed someone with an informed perspective on this.

Teri Dahle is a Blackfeet artist, photographer, and graphic designer. Her work depicts images of buffalo and horses against vibrant dreamscapes of teepees and plains landscapes. In addition to being a talented artist, Dahle is also an energetic advocate for bison. She recently stepped down as the Iinnii Initiative coordinator for the Blackfeet Confederacy of Tribes. The Initiative brings together the Kainai, Siksiika, Piikani, and Amskapipikuni tribes from both sides of the U.S.-Canadian border in a long-term bison restoration effort. The aim is "welcoming the Iiniiwa [buffalo spirit] home to again live among us as creator intended." The ultimate goal is to have bison grazing freely across tribal lands and migrating seasonally into the foothills of the Badger-Two Medicine Cultural District and Glacier National Park. A recent Blackfeet Tribal Business Council proclamation put the connection between cultural lands and bison in blunt terms: "The Badger-Two Medicine Traditional Cultural District is incomplete without the buffalo."

Dahle's role in the initiative was to connect people back to the buffalo and the buffalo back to the land. A steering committee of elders guided her in a series of steps to ensure the ways of the buffalo were properly reintegrated into tribal life. She created a four-day festival called "Iinnii Days" to celebrate the delivery of new animals from Elk Island. She organized the Iinnii Film Festival and Art Show in which people gathered to watch documentaries, share laughs, and talk about art into the night, their energy stoked by generous helpings of buffalo chili. She spent hours in the schools, working with teachers to get more information about buffalo into the curriculum. She mentored young interns to go out in the community to talk about the bison's return. "It's so important spiritually and mentally that we have those buffalo back," she told me. Ervin Carlson, the manager of the Blackfeet's bison and Dahle's boss, was also clear about the stakes: "These animals are culturally and spiritually connected to our people. . . . their homecoming will begin a healing of historical trauma to the Blackfeet people."

When I asked Dahle about the differences between the bison in the commercial herd and the newer bison arriving from Elk Island and American Prairie, she acknowledged the genetically pure animals had particular significance. The tribe valued the stories of Little Falcon Robe, the decades the bison spent in the Mission Valley, and their years in exile in Canada. Having these bison back was the proper culmination of a century-long odyssey.

But the more Dahl talked about the buffalo, the clearer it became their significance was not particularly rooted in genetics. In fact, she found all the genetic talk a bit of a distraction. "I went to a lot of meetings and different groups," she said. "Purity is a huge topic. It kind of pisses me off, to tell you the truth, because us as Native Americans, it is almost like saying, 'If you are not a true 100 percent full-blood, then you are not an Indian.' The same thing happens towards the buffalo." Blood quantum had always been a tool for discrimination.

Every bison was important, as Dahle saw it. "It doesn't follow that a buffalo which is 96 percent pure does not have value to us. Just because they are not 100 percent, they still look the same, act the same, have the same meaning. Why is this a big deal? Everything is the same." Dahle once asked a bison expert at the Ted Turner ranch if there had been any studies on the difference genetic purity makes. Researchers in North Dakota had found the loss of purity may have a small cost for the bison's resilience in bad weather. Other researchers had found a marginal reduction in weight among hybrid animals. But little else was different. The nutrition in the meat was the same, the ecological role was the same, the shape of the animal was the same.

I asked if her view was widely shared within the tribe. "I think it's widely shared amongst our people. They look out in the field and feel good they see buffalo out there. When they eat the meat, they are not saying, 'I wonder if this is 96 percent.' No, nobody cares. Not our people. They don't care."

Dahle is not blind to the importance of genes from a biological point of view. She knows genetic diversity is important to the long-term prospects of the herds. She is keen to keep bringing a mix of bison onto the reservation. She would like to see some of the bulls from Elk Island introduced to fertilize the commercial herd to create more diversity there. But while diversity mattered, a scientific standard of purity was far less important and even downright harmful.

Danny Kinka, the ecologist at American Prairie, was also open to a more expansive approach to bison genetics. When we were done talking about his work at American Prairie, we spent time philosophizing broadly about a range of topics in ecology. He made it clear he was now speaking outside of his professional role. On the one hand, he knows that conservationists need a way of being precise about the animal they are trying to save. Genetics can help with that. It can also be important in specific contexts like American Prairie's. But on the other hand, he thinks you can take it too far. "Speaking as an ecologist, why would this make a difference? Ecologically, I get it. But ranked against all the other stuff we are up against when it comes to rewilding and conservation science, that is the bottom of the list. There are too many other things to worry about. They are still an ecological engineer. They still behave like bison, and they are still doing ecological work there out in the field."

Caring about genetics over ecology role sometimes boiled down to whether you were a splitter or a lumper, said Kinka. Splitters are keen to define exactly what a species or subspecies *is*. Lumpers are more interested in what they *do*. Splitting had long been in the driving seat of conservation. "Natural science had a lot of splitters in it historically," Kinka said. "There is so much about subspecies in the early literature. But it seems like most people these days on the ecological side are more in the lumper category." Kinka also worried that obsessing over genetic identity is at odds with the dynamism of genomes. Specifying a particular genetic identity is like freezing one frame of a movie and thinking it is the only frame that matters. "It's like you want to put a pin in time. But ecology teaches you that everything changes all the time."

This sounded fine, but if the value of bison was not all about genes, what else counted? Speaking with Dahle made it clear the measure of the buffalo for the Blackfeet was as much about their place in Blackfeet stories as how they stacked up in genetic quantum. Buffalo mattered because of the role they played in tribal history. Stories about bison gave structure to tribal life. Not all of these stories were in the distant past. Dahle told me one of her goals was for tribal members to know more about the buffalo's recent return to the reservation. It was her great-uncle who first brought bison back to Blackfeet country in the 1970s. "He just dumped them out in the middle of the reservation," she chuckled. The buffalo tore down fences and got into haystacks, and generally made a nuisance of themselves. The tribe realized they had to secure dedicated prairie on which the bison could live.

Dahle told me about the different ways they managed this. One of them involved some quick thinking by a recent tribal chief. A few years back, Chief Earl Old Person had found himself driving around the reservation with the U.S. Secretary of the Interior in the car. It was just before the secretary's term ended, and she wanted to show goodwill toward the Blackfeet. In the middle of the drive, the secretary turned to Chief Old Person and said, "Well, if I could do anything for you, what would you like, Earl?"

Old Person was a quiet man. He did a lot of listening but had the speed of an antelope when required. At that moment, he and the secretary happened to be driving by a parcel of land taken from the tribe by the military during World War II. Old Person pointed a wrinkled finger out the window and said to the secretary, "How about you give us that land back. I'll give you some meat for it." The deal was done, and the Blackfeet gradually pieced together enough acres to house their buffalo, one gleam in an elder's eye at a time.

* * *

Just when I thought I had settled the question about the genetic purity of recovered wildlife in favor of relaxing the ideal, I had a conversation that demonstrated the dangers of generalizing. Tom MacDonald, manager of the Confederated Salish and Kootenai Tribes Fish and Wildlife Division—the reservation where Roy Bigcrane lives—was not as casual as Dahle about genes. MacDonald's agency has won an array of awards for its successful work in recovering wildlife, including trumpeter swans, grizzly bears, and native westslope cutthroat trout. The agency also participates in a long-distance comanagement arrangement for bison wandering out of Yellowstone National Park in winter.

After forty years working with wildlife for the tribe, MacDonald still thinks genetics count for a lot. He told me about tribal elders who rank the genetic inheritance of wildlife highly. They expressed concern to him about the authenticity of hatchery fish used to stock streams on the reservation. They cast doubt on the value of trees grown in greenhouses and, inevitably, questioned the quality of buffalo riddled with cattle genes.

"You can't do better than mother nature, especially when you are talking about genetic diversity and adaptation through time," MacDonald told me. When it comes to fish, the wildlife division does what it can to avoid hatcheries. "Our hatchery is the native stream," he said. With bison, it was the same. "A genetically pure bison is the goal," he assured me. MacDonald thinks being a purist has a practical benefit. "If you want to talk about an animal that will survive climate change, you want a wild buffalo. They are tougher than shit."

But even MacDonald admitted that sometimes chasing down the last gene is unnecessary. Some of the tribe's efforts to restore native trout rely on a technique known as genetic swamping. You just keep pumping good genes into a compromised system by introducing fish you know to be pure. The new fish start to swamp out the hybrids already in the creek. You can never turn the clock back entirely, but eventually the amount of hybrid genes in the fish becomes so small it is hardly worth worrying about.

* * *

Whatever the genetic future of the Blackfeet's buffalo herds, their recovery has opened up a wide swathe of opportunities for the tribe. Each year, a group of volunteers participates in an annual drive of the buffalo in the commercial herd. The buffalo aren't allowed off the reservation yet, and they have never set foot in the National Forest or the National Park to the west. Nevertheless, the three-day drive on horseback and ATVs gives the volunteers the feel of being around buffalo again. The Blackfeet Buffalo Program harvests a few animals each year and distributes the meat to the elderly and the sick. Since 2018, the tribe has also exercised its treaty rights to hunt buffalo around Yellowstone. The eighty permits for tribal members sell out within hours. Demand for buffalo meat on the reservation is growing thanks to the efforts of people like Dahle and Carlson.

The tribes of the Blackfeet Confederacy are working together to realize the possibilities for buffalo recovery. When Dahle was asked to write a business plan for buffalo on Kainai lands, she said it initially threw her for a loop. "It is really hard for me to write a business plan for something that is mental. The goodness that it brings, all the way around. The restoration of the land, the restoration of the people. Why put a dollar amount on it?" She wrote the plan anyway but told me that, even as she wrote it, she already knew money was not going to be a problem. "What's going to happen," she told the Kainai, "is these nonprofits are going to throw money at you. They are going to come running." Money, she said, should be the least of their concerns. "For 125 years," she told them, "buffalo have not walked across your grasslands. Think about that."

Lou Bruno also made the economic case to me. Bruno is a former president of the Montana Wilderness Association and cofounder of the Glacier-Two Medicine Alliance. "There's a ways to go," he told me over tea one winter's afternoon in a Missoula coffee shop. "But there is no doubt the possibility is there." Moose have returned in numbers to the Rocky Mountain Front. Grizzly bears and wolves are venturing farther and farther onto the plains. Swift fox have been reintroduced. Deer and elk numbers are healthy. Huge free-roaming herds of bison would help complete the picture. Three million visitors a year come through Glacier National Park just to the west of the Blackfeet Reservation. If a small portion continued onto Blackfeet lands to see wildlife, it could create a sustainable income stream for the tribe. According to Bruno, this is one of the few places in North America where a complete restoration of native animals remains possible.

America's national mammal has come a long way since Little Falcon Robe walked a handful of animals across the mountains a century and a half ago. But bison remain unique among wildlife. Nowhere are they allowed to roam completely free. Conservative legislators in states like Montana still do their best to restrict the bison's movement. Yellowstone bison are hazed at the park's borders, shot under harvest permits, and taken into quarantine by park managers. Bison are big animals, and they rarely respect the fences stitched across the prairie over the last century. When the dubious claim that bison are superspreaders of disease is thrown in, you find hesitancy about their return remaining high.

But the Great Plains are big. And some residents are looking for something more inspiring than the harrowing tales of decline they inherited from the last century. The fabric that joins people and bison is beginning to be rewoven. The genetic puzzles will continue to pulse from background to foreground in different minds at different times. But the lines between what's desirable from a genetic point of view and what's not are much fuzzier than I originally thought. Steppe bison bled into aurochs in Eurasia a hundred thousand years ago. Ancient herds moved back and forth across the Bering land bridge as the ice came and went. Plains bison took on cattle genes in North America when the species came back from the brink. People, wildlife, and the grassy contours of the northern plains have been mixed and then remixed over the ages. And for twelve decades now, as the politics has raged around them, bison numbers have continued to tick relentlessly upward.

RIVER ENGINEERS

1 HEALING DESTRUCTION

I was cold and wet. The rain came down in torrents, testing the waterproofing on my commercial fishing gear to its fullest. Around me, the mist hung thick on mountainsides draped in Sitka spruce and hemlock. Drips fell every few seconds from the bill of my cap. I kept my head down to prevent the cold rain from pouring down my neck. I also kept my head down to avoid a face full of fish semen.

It was spawning day at the Wally Noerenberg Hatchery in Prince William Sound. I was standing next to a cement channel half full of running water. Two jovial fish techs stood next to me, one of whom coached me on how to extract milt (the fluid containing sperm) from a mature salmon. It is a messy business. You start by grabbing a big male from a trough of anesthetized fish. With a good grip on the slippery body, you turn the fish on its back and run a thumb firmly down its belly, forcing sperm from an opening by its anal fin. The goal is to get the milt into buckets full of eggs extracted earlier from the females. If all goes well, the milky substance jets straight out of the fish and into the bucket. If it doesn't, the brim of your hat had better be low.

The fish techs put the fertilized eggs in trays stacked in a warehouse. Plastic nuggets simulate gravel on a creek bottom. Water from a reservoir above the hatchery constantly flushes the eggs to keep them clean and oxygenated. The cement floor amplifies all the sound, creating the impression you are standing in a waterfall. As winter progresses, two dark eyes will grow in each tiny pink pearl. The eggs slowly transform into something resembling a twitching matchstick in a thin, transparent sock. The matchstick has a small yellow sac attached to its abdomen. These odd-looking creatures are alevins, the first life stage of a new salmon.

Taking milt and eggs from hundreds of salmon is a challenging way to spend an afternoon. It's cold. The rain gear is uncomfortable. Everything smells of water and slime. But it doesn't take much imagination to realize something remarkable is going on. Despite the absurdity of spraying sperm from an unconscious fish into a bucket of gelatinous spawn, you are a witness to the culmination of thousands of years of evolution. The banter of college-age fisheries students and the scrape of buckets against cold cement masks a breathtaking act of creation.

Hatcheries like this one in Prince William Sound provide a glimpse of the astonishing productivity of fish. One mating pair can hatch thousands of young alevin. When you add this to the hardwired drive of salmon to spawn, you get an eye-popping potential for resurgence. This is true across much of the fish world. The sheer fecundity of fish makes possible rapid recoveries in rivers restored to their former health.

The story of river recoveries is both inspiring and exasperating. It is always complicated. River restorations involve laborious clean-ups, spectacular dam removals, and controversial hatcheries. They also involve a whole heap of arguing. When they go right, the rewards can be astonishing feats of wildlife recovery. But to go right, they demand an honest answer to a pressing question. It's a question that a species inclined toward great feats of technical manipulation finds difficult to ask, let alone to answer. The question is this: can humanity reign in its insatiable desire to engineer the flow of water?

* * *

A day and a half of consistent rain: that's all it took to turn the lethargic assembly of fish at the mouth of the creek into a full-on salmon run. For nearly two weeks, I had watched a growing throng of pink and chum salmon swim lazy circles where the creek emptied into the sea. A few recreational fishermen cast lines into the pattern, but the fish weren't biting. Summer was fast turning to fall, and it was too late in their lifecycle for the salmon to have any appetite.

Three weeks without rain had reduced the creek to a trickle. Only a handful of fish had made it through the shallow water to the footbridge two hundred yards upstream. I watched from the bridge as a pair of chum salmon hovered over a slight depression in the streambed. At a moment that felt right, the female rolled onto her side and beat her tail rapidly against the gravel, creating a small depression for her eggs. The male quivered in anticipation, but no eggs came. With the water level so low, nobody seemed ready to expend much energy.

Then everything suddenly changed. The gray skies cracked open, and rain fell in long, straight rods. The creek rose six inches overnight. The Tongass became a giant wet sponge leaching delicate aromas from its peaty soils. An ancient spark fired deep inside each fish's brain. Dull migratory instincts became razor sharp again. Salmon that had spent eighteen months or more at sea could suddenly smell their way home.

Over the next few weeks, thousands of salmon would ascend the creek and deposit their eggs in redds (depressions) dug out by the females. The adults would sniff out the same pools in which they hatched several years before. Their work accomplished, the spent fish would lie down in the shallows to die. In a noble last act, their decaying bodies would provide nutrients for the young that hatched a few months later.

The scene repeats itself in thousands of streams up and down the Pacific coasts of North America and Eurasia. It is a stirring example of nature's profusion. Alaska is one of a handful of places where the salmon runs that once filled rivers across the Northern Hemisphere still show strength. Storms barreling in off the Pacific soak the coast with 150 inches of rain a year. Deep forests protect many of the headwaters. Glistening cascades of clear water dump into the ocean from a million boulder-strewn creeks. The greater the flow of water, the more fat the fish need to ascend to the spawning grounds. The combination of muscle and fat on Alaska's fish gives them a prized oily texture, one Native people in the region have cherished for millennia.

Further south, things haven't gone so well. From British Columbia to the central California coast, the runs of all five Pacific salmon species have been decimated by what fisheries managers call "the four H's"—harvest, habitat, hatcheries, and hydropower. Hydropower is particularly devastating. Salmon may be able to wait out the occasional dry spell at a river mouth, but they have no answer to a concrete wall blocking their passage upstream. If American settlers had wanted to kill off the salmon they found in virtually every northern river and creek they encountered, there would have been no more effective way than to block all the free-flowing rivers with dams.

* * *

Americans spent more than two hundred years proving they knew how to plug a river. They built an average of one large dam every day after the Declaration of Independence was signed. The dams were useful. They provided power, stored water for irrigation, and offered a line of defense against flooding. But they also had a catastrophic effect on migratory fish. Adults heading for spawning grounds above the dams were completely cut off. Salmon, steelhead trout, shad, striped bass, smelt, sturgeon, river herring, and lamprey all suffered steep declines.

A century and a half ago, the northwestern states appeared to acknowledge the problem. Written into the Oregon Territorial Constitution of 1848 was this sentence:

> The rivers and streams of water in said territory of Oregon in which salmon are found or to which they resort shall not be obstructed by dam or otherwise, unless such dams or obstructions are so constructed to allow salmon to pass freely up and down such rivers and streams.

Oregon's law had distinguished precedent. In twelfth-century England, Richard the Lionheart issued a decree mandating that all dams should have gaps the size of a well-fed pig standing sideways. Not long afterward, Britain's Magna Carta ordered the king to remove fish weirs to allow salmon passage. By the time Oregon came on board, it was well-known dams posed a threat to fish. California passed a law similar to Oregon's in 1852. Washington state followed in 1890.

The laws may have been on the books, but dam builders often simply ignored them. There was too much money to be made. Legislators prostrated themselves to powerful industries, turning a blind eye to what the statutes said. Manifest destiny, according to influential nineteenth-century journalist John O'Sullivan, meant all American lands were "allotted by Providence for the free development of our yearly multiplying millions." Guilt or caution was unpatriotic. Hydropower was an effective way to spur development, and the turbines wouldn't spin without dams.

Fish struggled even when dam builders did comply with the law. Stepped pools of flowing water known as *fish ladders*, built to help fish bypass a dam, are only partially effective. They slow fish down and can be too awkward for species like sturgeon and lamprey to negotiate. Fish in line for a ladder are forced to wait at the base of the dam where predators make the most of the easy buffet. When they do make it up the ladder, some exhausted fish are immediately flushed back down by the force of water.

The problems multiply in the spring when the juveniles head toward the sea. A salmon smolt journeying downriver wants to feel a current. Slack reservoir water makes navigation difficult. Migration times through dam-choked rivers like the Columbia and the Klamath slow from a few days to several weeks. Varying water temperatures in the reservoirs add to the confusion. Algae block the fishes' gills.

When they finally arrive at a dam, some fish are pulled over the bruising spillways. Others are sucked into penstocks (the pipes that run steeply from the reservoir down to the powerhouse), where the pressure can be fatal. The hydropower turbines at the bottom act like Cuisinarts on anything sucked through them. The survivors' bodies get stressed by fluctuations in dissolved oxygen levels as they cross each impoundment.

In an effort to improve the fish's odds, in 1955 the state of Washington experimented with shipping 200,000 chinook salmon around the Bonneville Dam on a barge, releasing them downstream near the river mouth.

Today thirty million juvenile fish are moved around the Columbia dams using barges, trucks, helicopters, and even water cannons. Avoiding mortality at a dam site is obviously good. But a fish that catches a ride in a helicopter or barge doesn't learn much about the river of its birth.

The combination of dams and reservoirs can kill more than half the salmon trying to cross them. A comprehensive 2020 report found migratory fish around the world have plummeted by three-quarters in the last fifty years thanks in large part to dams. The Columbia River used to see sixteen million salmon each season. Its four hundred dams have reduced salmon and steelhead to a million. The story is similar up and down the West Coast. The Klamath River canneries packed seventeen thousand fish a day before the era of big dams. Runs are now less than a tenth of that. Close to 30 percent of the Pacific salmon runs south of British Columbia are extinct. A third of the remainder are endangered.

By many accounts, the future of migratory fish looks dismal. Climate change is warming creeks and changing ocean conditions, making the odds of survival even longer. But eight hundred years after the Magna Carta, the recognition that dams kill fish is beginning to sink in. And this creates a sliver of hope. A single salmon can produce thousands of eggs. A single striped bass can lay more than three million. When a dam comes down and good spawning habitat becomes available, migratory fish can come back strong. The potential for recovery is explosive. And explosions, it turns out, are sometimes what is involved.

* * *

Robert Elofson of the Lower Elwha Klallam Tribe stopped midstride and turned toward me. "Tell me what Boulder Creek looked like," he asked, his voice lifted with excitement.

"You mean the creek itself? Well, it comes out of the forest about ten feet wide and cuts through the sediments toward . . . ," I started.

"No, no. I know what the creek looks like. What about the fish? What were the fish like?"

Elofson is a short man with a round face and glasses. He smiles frequently and, I had learned over the last couple of hours, is exceptionally warm to strangers. Until earlier that day, we had never met. I had called a few weeks before and told the former director of river restoration I hoped to come to the Lower Elwha Klallam Reservation to learn about the impacts of dam removal on the Elwha River. He told me he would be happy to help out. When I showed up that morning, he welcomed me into his office at the tribal Natural Resources building as an old friend and volunteered to show me a few sights.

When he asked about Boulder Creek, we were standing a few hundred feet from the mouth of the river on a beach belonging to the tribe. We had come there so Elofson could show me how sediments stacked up behind the dam had rebuilt the estuary when the newly freed river flushed them out. I had been telling him about my hike the previous day up the Elwha. I wanted to see salmon spawning above where the upper dam had stood. Mike McHenry, the tribe's habitat program manager, told me Boulder Creek was the place to go.

The former dam site was about three and a half miles from the trailhead. The tang of fall was in the air as I set out on my bike up the abandoned Park Service road. Bigleaf maples marching up the slopes on each side of the river valley had dropped leaves the size of dinner plates on the road. The Elwha ran in emerald braids nearby. The blacktop climbed gently as it headed toward the heart of the park. After stashing my bike at the dam site, it was another two miles along a rough track through thick alders to Boulder Creek. Elofson used to dive for sea cucumbers and still fished commercially for crab. He showed me with pride some new crab pots in the back of his truck. Despite still being active on the water, at sixty-eight, his legs weren't up for the trip to Boulder Creek any more.

His face softened when I told him about the forty chinook salmon I had seen milling around where the creek met the main stem of the Elwha. His eyes lost focus for a few seconds, and his mouth cracked a smile as he conjured up images of what I had seen. It was the second time that day I had glimpsed this change in his face. Standing above a cement raceway in the tribe's gleaming hatchery a couple of hours earlier, Elofson had become transfixed by the adult coho salmon swimming patterns beneath us. "Look at those beautiful fish," he said, as much to the surrounding forest as to me. "... Beautiful fish." We stood there for a few seconds in silence. His reaction to the salmon made something completely clear. For a Lower Elwha Klallam tribal member, a salmon in the Elwha is far more than just a fish.

* * *

The Elwha Dam stood tall on Washington's Olympic Peninsula for close to a century. Constructed in 1912, it helped spur development of the region's timber economy. The Olympic Power Company used the dam to lure pulp mills to Port Angeles with the promise of cheap power. The hunger for development meant that Thomas Aldwell, the Canadian businessman who built the dam, simply didn't bother getting a license. Legislators turned a blind eye. When the Glines Canyon dam joined the Elwha a decade later, seventy miles of prime spawning habitat became inaccessible to migrating fish.

As many as 400,000 salmon returned to the Elwha River the year after the dam went up. They spent weeks launching themselves against the cement until they were bloodied and bruised. Elofson's grandmother walked half a mile from her home to the new dam. She told her grandchildren she watched salmon "jumping, jumping, jumping, trying to get past" and then went back home and cried for hours. The Elwha's fish were famously long. Elofson told me of fish slung over grown men's shoulders with tails scraping the ground. When the dams blocked the river, the hundred-pound "hogs" had nowhere to go. Over the next few decades, the number of fish plummeted from nearly a half million to three thousand. The few survivors spawned in the five miles of habitat still available below the dam.

The Lower Elwha Klallam tribe felt massively betrayed from the start. Salmon were woven into the seasons of their lives. The fish were once so plentiful elders used wheelbarrows to bring the catch home. The old delta at the Elwha's mouth had been stiff with clams and flounder. Sediments carried downriver created habitat for the sand lance and smelt on which the young salmon fed. When the dams went in, the tribe lost not only the food that sustained them for centuries but also a set of ancient ties to their surroundings. The reservoir behind the Elwha Dam flooded a pair of rocks at the tribe's creation site. The rising waters obliterated a tribal village. Ceremonies to honor the fish's annual return faded away. To add insult to injury, part of the dam blew out in its first year, damaging homes downstream and littering salmon all over the fields. Tribal chairperson Frank Bennett summed up how this felt: "I guess they don't care if a few Indians drown." The tribe could see what was at stake. It was not only their traditions and livelihoods in jeopardy. It was their lives.

In 1986, the Lower Elwha Klallam filed a motion to remove the dams. The Federal Energy Regulatory Commission, the authority responsible for permitting hydropower, were debating a relicensing application from the dam's owners, the James River Corporation. New laws required dam owners to do much more to protect fish and wildlife. The relicensing was going to be expensive. The tribe argued removal was a better option on safety grounds. The Elwha Dam had leaked since it was built. The concrete had deteriorated. Earthquakes in the Pacific ring of fire threatened to shake both the Elwha and the Glines Canyon dams to the ground.

The tribe also argued for removal on cultural grounds. The destruction of the salmon run had caused them direct, personal harm. Tribal chairperson at the time Dennis Sullivan put it this way: "We are protectors of the salmon. Salmon and us are like family to each other. We need each other."

Tribal elder Bea Charles went to a Senate Committee on Energy and Natural Resources hearing in DC in 1992 and forced the assembled senators to pay attention. "I remember how the fish runs were," she testified:

> It was just ripples of salmon going up, and they were big salmon. I remember. I saw it, and I knew that it was there. . . . Our Creator gave us this fish to live on, and it was rich and an abundance of fish. It was given to us, and we cherished it and we respected it. We never got more than we could use. We used it and every bit of it. . . . It was our culture and our heritage.

The argument for tribal justice never held much weight when the dams were built. But public opinion had changed. Residents of the Pacific

Northwest had woken up to the wrongs unleashed by state and federal governments on Indigenous people. A decision by Judge George Boldt in 1974 affirmed that Northwest tribes possessed treaty rights to 50 percent of the fish on their traditional fishing grounds. The collapse of the salmon runs meant this hard-fought right amounted to 50 percent of nothing. Even nontribal members now understood this.

Many of the reasons to build the dams in the first place no longer made sense. The hydroelectricity was no longer needed in Port Angeles. The dams provided little in the way of flood control and zero irrigation benefits. Remnants of the town's industrial past certainly remained in evidence. A pulp mill at the base of Ediz Spit still operated, and a shipyard still built yachts. But the regional economy was no longer hitched to lumber. Recreation and tourism were growing, and 83 percent of the Elwha's watershed was inside Olympic National Park, a World Heritage Site. Growing numbers of people arrived in RVs and on touring bikes each year to admire the steep peaks and glistening waterfalls. Blocking a river that flowed out of a national park and through a tribal reservation now looked criminally wrong.

The U.S. Department of Interior studied the case for removal at the tribe's request and issued a ruling in support. The fate of the dams was sealed. It would be the largest dam removal in the world to date. The Park Service sat down with the city of Port Angeles, the tribe, and Washington state to work out the details. The tribe asked that no visible sign of the lower dam should be left. The Elwha River Restoration Act made provision for transfer of some of the recovered land for tribal use.

On a clear September day in 2011, tribal members joined dignitaries from Washington, D.C., and across the region to watch an excavator growl its way toward the dam with its bucket raised. When the bucket crashed down on the concrete structure, tribal members rose to their feet and beat ceremonial drums. By the summer of 2012, the Elwha Dam was gone. The Glines Canyon Dam followed in 2014, its last cement anchors dynamited from the bedrock by demolition crews.

Within weeks, steelhead trout spawned above the former dam site. In early fall, chinook salmon migrated past where the Elwha Dam had stood. The Lower Elwha Klallam held their first ceremony to welcome the salmon home in a century. A recovery had begun.

* * *

The species driving the removal of the tiny Rattlesnake Creek Dam outside of Missoula, Montana, are a little less mighty than the Elwha's salmon. But the bull trout is just as threatened as its coastal cousin. These voracious predators move up and down mountain creeks and need cold, clear water to spawn. The overachievers among them migrate 150 miles to find the perfect spot. Although the bull trout in Rattlesnake Creek are not trying to reach the ocean, dams have been a disaster for them. Like salmon, they have no answer to a vertical cement wall blocking their way home. Obstructed rivers, warming water, and deteriorating habitat have pushed bull trout to the brink.

The dam blocking Rattlesnake Creek was no longer required for water storage. It didn't do much for flood control, and every year it needed a little more maintenance. Removing it would reconnect twenty-six miles of bull trout habitat, much of it protected within the Rattlesnake Wilderness and National Recreation Areas. It was a perfect opportunity to do something for fish.

When I arrived at the removal site, Rob Roberts of Trout Unlimited was striding around with a clipboard, feeling, I suspect, more than a little pride. Roberts is a rugged-looking forty-year-old who lives a couple of miles downstream from the dam. He wore a yellow hard hat, a high visibility vest, and rubber boots. Two days of stubble shadowed the lower sections of a chiseled face. Today was the day the heavy equipment had arrived to start leveling the dam. There was a lot for Roberts to figure out.

One of the first orders of business was to make sure I didn't get mown down by a ten-ton dump truck. Roberts grabbed me a vest and hard hat from the back of his rig and steered me toward a quieter part of the work site. He pulled out his maps and plans and started to explain.

Although plenty of cement and retaining walls remained visible, three-quarters of the dam was already under siege. Yellow earthmovers had scraped the top off the huge berm of earth that formed 80 percent of the dam. They were spreading the material around the site to sculpt the floodplain that would take its place. Roberts showed me drawings of the future river, raising his voice over the grind of diesel engines and the crack of large rocks being dropped into the beds of the trucks.

If Roberts felt any pride at the work taking place in front of us, it was surely justified. When the city took ownership of the dam from the local water company a few years back, Roberts saw a golden opportunity to restore a creek with headwaters in an alpine wilderness. The fact the project was in his own back yard made it all the more personal. Missoula threw in a small amount of cash, and Roberts and Trout Unlimited raised much of the rest of the $1.5 million price tag.

We stood on the dam's western edge near the old fish ladder. We could see the ladder's dried-out remains ascending toward us from the creek below. The dam owners added the ladder as an afterthought, sixty-five years after the dam was installed. The fish swam up the ladder toward the reservoir but then had to navigate a dark, corrugated pipe passing through the middle of the earthen barrier. Biologists noticed the native trout made it to the top of the ladder but hesitated before entering seventy-five feet of dark tunnel. Engineers decided to add four vertical light tubes through the dam to provide daylight to the pipe. The tubes, now bent and exposed, tracked the route of the fish tunnel. It must have been one of the only dams in the world where fish had their own custom-built skylights.

Roberts explained to me the many wins the dam removal offered. "First, the city gets out from their annual maintenance costs. Second, they lose the worries around public safety caused by the decrepit cement spillway. Then there are the ecological benefits for fish, bears, moose, and other animals that will use the floodplain. Finally, the public gets new trails as well as opportunities to do citizen science." It sounded great, but I still wondered how an organization like Trout Unlimited justified raising this kind of money to save one small creek. "It's bragging rights," Roberts said with a smile. "This a high-profile project on the edge of a big city." He paused for a second: "It's also just the right thing to do."

Watching the giant backhoes move rock was a good reminder it can take serious money and commitment to correct past decisions. It's one thing to learn a lesson. It's another to put diesel in the tank to undo your mistakes. Skilled heavy-equipment operators, snarling engines, and half a dozen crew in hard hats were revitalizing a wasted creek before my eyes. You could see a different future emerging as the components of a healthy ecosystem were put back into place. This wasn't just about fish. The spinoffs radiated outward—fish, moose, recreation, safety, citizen science. The backhoes were spinning webs of possibility.

As if to drive home what I was thinking, Roberts told me there was also a public art component to the project. A local nonprofit had partnered with the city and Trout Unlimited to offer residencies for artists interested in rivers, land, and restoration. This was too interesting an element of the story to pass up. It didn't take long to track down one of the artists.

* * *

Amanda Bielby has a family interest in dams. Her father was a construction foreman who built access roads to dams going up throughout the west. Like many long-term construction workers, her dad suffered health consequences from his work and now was in a period of decline. The chance to witness up close the type of work her father had done was a poignant opportunity.

Bielby was in the process of completing her own recovery. Her previous job involved restoring artwork on historic buildings. Days spent recreating elaborate ceilings while maneuvering around scaffolds had taken its toll. Her back and her arms ached. Her tendons were permanently inflamed. She needed to recover her physical health. So Bielby quit her job and started focusing on her creative interests.

Rehabilitation was on her mind for other reasons too. She and her husband had just started a restoration project of their own. They recently bought some damaged riverfront property thirty miles away on the Clark Fork River. They planned to use it for hunting and camping, but the site needed some care. They picked up trash and tidied the old shop space. They spent weekends planting wild rose, dogwood, and juniper to stabilize the eroding river banks. The place was starting to come back into shape.

I met Bielby in her backyard art studio. She greeted me around noon wearing a smock and the weary smile of someone who has already put in several hours of work. Every square inch of wall space was covered by picture frames, display boards, and pots of paint. Carving tools and dozens of art pieces in various stages of completion covered the flat surfaces. A computer and a tiny office space occupied part of one wall. Her hands were covered in dust from moving the heavy sacks of filler used in her ceramic molds.

I asked Bielby why a dam removal needed an artist. "I think you have to show people the 'why' of what they are doing," she replied. "You need to show there is a real purpose for taking things back to what they were." The why had many layers. Artists come at dam removals from an entirely different angle from fish biologists and city planners. Having an artist's eye on the site would get beneath the bull trout recovery goals, the decisions about twisted rebar, and debates over public access.

Bielby's artistic style brings attention to small elements of the landscape. She finds patterns in rocks or fish scales, which she reproduces in larger, stylized form. Natural color and carefully designed inlays invoke botanical and geological themes. She takes the idea of renewal seriously. She buys her canvases from a local nonprofit selling reusable construction waste. Her colors are saved from previous painting jobs. I learned that a deep commitment to recovery permeates all her work.

She told me about her goals for the residency. "Being outdoors, doing my smaller pieces is all physically healing," she said. As the Elwha had already made clear, a dam removal can rebuild more than just a fish run. Bielby intended to look closely at how the workers fared at the dam site as the weeks went by, remembering her father. She had decided to change her diet to include more plants and eliminate store-bought meats as a way of doubling down on her health.

"Going back to roots helps," she said. "I want to explore how I can intertwine my passion, my health, and my work all in one." River restorations are a way of reconnecting things, Bielby was telling me. Fish are just one element of a much larger web of recovery.

* * *

Back on the Elwha, the river needed time. Thirty-four million cubic yards of sediment trapped behind the dams had washed downriver after demolition. The sediment pulse made conditions worse before they got better. It smothered fish eggs and suffocated alevin. It destroyed kelp beds at the river mouth. The water's turbidity was worse than it had been in a century. The river also became volatile again during spring rains. A Park Service road washed out, and a campground became unusable. When I asked Elofson if the flooding alarmed tribal members on the reservation downstream, he scoffed. "The river has always flooded," he said. "This is nothing new." Elofson thought the loss of the road was a heartening sign the river was alive again.

Everyone knew the salmon would need time to adjust. Fish with genes suited to the Elwha's upper reaches had not ascended the river in a hundred years. But within two and a half years, nine runs of anadromous fish—five runs of pacific salmon, two steelhead runs, a Pacific lamprey run, and a bull trout run—had all resumed migrations from the sea to fresh water above where the Elwha Dam had stood. Within five years, chinook salmon and steelhead trout were spawning in over thirty additional miles of river. The tribal hatchery occasionally placed fish in various tributaries to speed things up. Future generations of salmon would in all probability find their way to every accessible riffle. Biologists are hopeful that up to 400,000 salmon will eventually swim their way up a revitalized Elwha to reach spawning grounds abandoned for a century.

In the meantime, the river has experienced a cascade of effects from dam removal. A welter of wild beings swam, crawled, and swooped their way back. Summer steelhead, landlocked for generations, resumed their migrations to the ocean. The complex of sandbars, ponds, and riffles created by the free-flowing waters enhanced insect abundance, benefitting fish, bats, and birds. Tons of sediment flushed out the mouth of the Elwha, where it created the new estuary Elofson showed me. With the cobble replaced by sand and gravel, the estuary quickly attracted the shellfish his grandmother used to harvest. Juvenile coho heading to the ocean were now twice the size recorded before dam removal.

The growing number of fish surging upriver became nature's own fertilizer. A large salmon contains 130 grams of nitrogen, twenty grams of phosphorus, and more than twenty thousand kilojoules of energy. This nutrient-laden package propels itself dozens of miles inland. When the fish dies, bears, eagles, and raccoons grab its decaying body and haul it into the forest. A single bear can take seven hundred fish into the woods over a season. The fine-dining bears often eat only the tastiest parts—the brains, eggs, and stomachs—leaving the rest of the salmon uneaten on the forest floor. The nutrients are taken up by roots of Sitka spruce, red cedar, Douglas fir, and hemlock. The excrement of the scavengers adds to the life-giving flush.

More than 70 percent of the nitrogen in a coastal forest comes from fish. Nutrients from salmon are detectable in the uppermost needles of two-hundred-foot conifers. In years with large salmon runs, older trees have bigger growth rings, creating a living data set of fish abundance. In the past, a river like the Elwha might be littered with three tons of salmon per mile at the end of spawning season. From bridges and riverbanks, I looked down to see the spawned-out carcasses of salmon starting to dot the banks again.

The salmon get a considerable return on their investment in the forest. The thick canopy of trees shades the streams and keeps the water cool. Their roots stabilize the stream banks and prevent soil from entering the water. Deep carpets of moss slowly release the rainfall, keeping the creek clean and preventing the salmon eggs from suffocating. Insects crawling through the undergrowth fall into the water and become food for young fish. When the trees die, their trunks topple into the creeks and create deep spots that become places for salmon to rest. It took thousands of years for the relationship between salmon and trees to develop. The result is an intricate, supportive system, poised to promote life.

The 137 species that directly benefit from salmon in the Pacific Northwest range from killer whales and the all-white spirit bear to mink, shrews, and salamanders. Jays, ravens, eagles, and crows, as well as many kinds of gulls and grebes, all feed on the carcasses. One of the most surprising beneficiaries is a songbird that bobs up and down on damp rocks in the middle of a creek. The dipper is America's only aquatic songbird. They can dive up to twenty-feet to the river bottom and mosey along looking for insects and salmon eggs using their wings as paddles. Dippers on the Elwha have become bigger since salmon returned. They are also twenty times more likely to raise two broods of chicks in a season than they were before restoration. When you give an ecosystem its fish back, everything ratchets up a notch.

The river is still a work in progress with several decades of recovery ahead. When I asked Elofson what else he wanted to see happen, he raised the issue of land. He hadn't forgotten that the congressional bill that authorized dam removal also promised land transfers back to the tribe. As well as land at the dam site itself, Elofson was interested in the possibility of a harbor for tribal fisherman nearer the reservation so they did not have to motor from Port Angeles every time they wanted to fish.

As the salmon recovered, Elofson wanted to see tribal fishing traditions return. "Opening the river back up for a subsistence fishery would be a start," he said. "Maybe we can have a dipnet fishery for cohos at some point." The tribe wanted the shellfish harvest to grow, both for their own health and to restore their traditions. Elofson also had a more philosophical goal: "I just want to see everything back the way it should be—the fish, the crabs, the clams." Harvesting food from the ocean was in Elofson's blood. These days, he survived mostly the modern way. "I buy most of my food from the grocery store," he confessed with a wink. But taking advantage of the seasonal catches was integral to his identity. The prospect of more tribal members eating traditional foods kept him motivated.

Things on the Elwha are heading in the right direction. The speed of the steelhead recovery has surprised everyone. Salmon numbers are slowly climbing as fish spawn farther up the river. Insect abundance is increasing. Orcas have been spotted patrolling the mouth of the river now that chinook salmon are returning. In the words of *Seattle*

Post-Intelligencer journalist Lynda Mapes, who covered the removal, the Elwha is "a river reborn."

Engineering rivers like the Elwha back to health provides many lessons in ecology. But it also offers some valuable instruction in humility. Blocking hundreds of rivers without regard for the ecological effects did not turn out to be as smart as it first seemed. Removing dams can be a way to correct this mistake. Although dam removals are often an essential starting point, even the most ardent removal enthusiasts concede that dam building still has its place. Sometimes, to re-create a river in which fish can flourish, you need a few dams scattered around. At the same time as ecologists are arguing for the removal of dams in some places, they are equally excited about the building of dams in others. But the builders they celebrate are not human engineers thinking of kilowatt hours and summer water storage but furry, orange-toothed rodents thinking about succulent willows and a safe place to raise their young. An unexpected ally in the return of migratory fish is a growing population of beavers. And beavers have a thing or two to teach about how to restore a river.

2 SO SALMON COULD LEAP

Mike Kustudia beamed from beneath his Montana Fish, Wildlife, and Parks hat on a warm morning in June. He was leaning on a digging tool known as a *Pulaski* while he waited for his trail crew to assemble near the old Milltown Dam site. Their task for the day was to reroute a small section of footpath that snaked its way across the floodplain. The weather was too nice not to spend a few minutes enjoying what the rebirth of a river looks like. This is especially true when you are standing where several feet of toxic sludge used to clog the bottom of a reservoir.

Kustudia is the manager of Milltown State Park, a 645-acre parcel of land at the confluence of two of Montana's most storied rivers. To the east, the Clark Fork, named after William Clark (of Lewis and Clark expedition fame), meanders in from the upstream mining towns of Butte and Anaconda. To the north, the Blackfoot runs crisp and clear for seventy-five miles before joining the Clark Fork in front of us. The Blackfoot is the focus of Norman McLean's *A River Runs through It*, a book whose 1992 adaptation into a movie made Brad Pitt famous.

The corridor in which the Blackfoot flows is known to Indigenous people as the Road to the Buffalo. This part of the Columbia River system never held salmon. But the Salish word for the confluence is *Ṇaaycčstm*, "place of big bull trout." The Nez Pierce, Salish, and Kootenai people used to journey up the river corridor to hunt bison over the continental divide beyond the river's headwaters. The confluence had long been a crossroads for trade and conviviality.

Kustudia trained as a journalist at the nearby University of Montana. After a brief career in newspapers, he spent two years in the Peace Corps in the Dominican Republic before returning to Montana for an environmental studies degree. Kicking around the nonprofit sector afterward, Kustudia volunteered on a citizen advisory committee formed for the removal of the Milltown Dam. Mining waste from Butte and Anaconda had turned the riverbed toxic for 130 miles. Thousands of tons of contaminated sediments piled up behind the dam, which led to arsenic and cadmium filtering into drinking water downstream. After thirty years of wrangling, a consent decree authorized removal of the dam and its sediments. Kustudia was awarded the job as the point person for a new state park to be built at the removal site. His easy smile and effortless charm made him the right person to lead the transformation of a poisoned industrial site into a community gathering point.

The position was particularly appealing because it connected Kustudia to his roots. His grandfather worked for forty years at the lumber mill at the confluence. His mother grew up just across the Blackfoot in Piltzville, a blue-collar community built by the timber industry. For decades, the mill offered the best-paying jobs around. The dam and its reservoir were all the scenery anyone could remember.

I asked Kustudia how personal the work felt. "Sometimes I think about my grandma," he said. "She never liked the idea of the dam coming out, just like a lot of the older segment of the community here. I do wish she could see it today." The mill site has been redeveloped for light industry. It now hosts more than a dozen businesses, including a brewery and a spectacular outdoor amphitheater on the river bank. Locals ride bikes and walk dogs along new trails. A number of the old-timers have stopped Kustudia on the street to thank him for his work.

Today the Clark Fork sparkled in front of us as it flowed freely past Piltzville and Milltown. Side channels and overflow ponds dotted the new floodplain. Every few feet, another willow or alder reached skyward from an expanding network of underground rhizomes. There was nothing left of the imposing cement structure that used to block the river a few feet from where we stood. There were, however, other dams on the site. It was those I wanted to see.

Kustudia leaned on his Pulaski and pointed me upriver toward a finger of trees that reached out from the hillside into the valley bottom. "Follow the trail toward where it turns to the river, and you'll see a fenced-off area. You can't miss them." He looked down at my boots. "You'll get your feet wet."

I walked along the old railroad bed past thick carpets of penstemon, flax, nodding onion, and woods' rose. A few hundred yards upstream, I turned toward the river and squelched through grasses soaked by high spring water. A can of pepper spray filled the back pocket of my jeans. Kustudia had told me to watch out for bears and moose bedded down in the willows.

The restored floodplain thrummed with life. Dragonfly wingbeats cracked the warm air. Deer and elk prints stippled the mud. The trill of red-winged blackbirds perched on gyrating bulrushes amplified the sense of abundance. I hopped across a seep and balanced my way over a damp log before finally finding what I was looking for. A scruffy barrier of mud and sticks three feet high stretched from one bank of a side channel to the other—a picture-perfect beaver dam.

* * *

Humans aside, it is unlikely any species shaped the landscapes of North America and Eurasia over the last few million years more than the beaver. Ben Goldfarb, author of *Eager: The Surprising Secret Life of Beavers and Why They Matter*, has suggested—with a nod to the beaver's Latin name—that before there was ever an Anthropocene, there was a Castorocene. The landscape had the signature of beavers stenciled all over it. Beavers altered the flow of billions of tons of water, shaping vegetation on a continental scale. An animal that controls water controls life, determining where other species can live. This gave the beaver outsized powers. As a result, they occupy a central place in the mythology of the Iroquois,

Tlingit, Athabascan, and Blackfeet people. When water is the currency, beavers control the cash flow.

The near total eradication of beavers from North America by the end of the nineteenth century ranks alongside the killings of Plains bison and the passenger pigeon as one of the most merciless slaughters of animals in human history. Nobody knows exactly how many *Castor canadensis* swam, waddled, and gnawed their way across North America before settlers arrived. The naturalist Ernest Thompson Seton estimated as many as 400 million. Placed nose to tail, this would make a line of beavers circling the earth twelve times.

Unlike the pigeon and the bison, the killing of beavers had nothing to do with food or cultural suppression. It was entirely about money. Rivers full of beavers were an untapped treasure trove to white men exploring the continent's interior. Their densely packed fur was perfect for felt hats. North American beavers helped meet an insatiable demand in Europe after supplies from Russia dried up. Beavers became known as "hairy banknotes." Merchant empires based on beaver skins generated tremendous wealth for traders. Today Beaver Street in Manhattan leads straight to the doors of the New York Stock Exchange. The seal of New York City boasts two beavers. The Hudson Bay Company's coat of arms sports four.

The promise of beaver skins lured French and English trappers deeper and deeper into country unseen by white people. It led to bitter conflicts and unlikely alliances among colonial powers, Native Americans, and mountain men. Beavers also helped create the mythology of an inexhaustible land that attracted wave after wave of European immigrants across the ocean.

Belief in the beaver's inexhaustibility quickly proved a mistake. Hunting and trapping crashed the population. Some Native American tribes assisted in the slaughter, trading beaver skins for pots, guns, and beads. The number of beavers plummeted from tens of millions to a hundred thousand. By 1841, a trapper named Osborne Russell suggested "it was time for the white man to leave the mountains." Their quarry was almost gone.

The sole reason beavers survived in North America is that the fashion for hats abruptly changed. In the mid-1800s, Chinese silk became popular in London and Boston, and the market for beaver fur dropped through the floor. Trading companies turned elsewhere for profits, which gave beavers a desperately needed lifeline.

The remaining beavers were not completely off the hook. They continued to be persecuted even after their furs stopped disappearing to Europe in the holds of merchant ships. Farmers and homesteaders didn't like their crops getting flooded when a beaver dammed a creek. The railroad companies resented clearing downed trees and repairing flooded track. Settlers hungry to scratch their own designs on the landscape wanted beavers dead.

As beavers declined, the wooden infrastructure assembled by generations of castorid dam builders started to decay. Beavers were no longer around to do maintenance. Blown-out dams sent sediments flooding downriver. The pulses of nutrients can still be detected in estuaries today. Vegetation changed as flood plains dried out. The continent's ecology shifted, aided by settlers keen to make the land more suitable for agriculture. The landscape—once so full of ponds, marshes, and floodplains that it was called "beaverland"—dehydrated rapidly. A diminishing portion of the continent now felt the ebb and flow of freshwater.

The loss of the beaver sparked a cascade of ecological effects. Beavers create a complex mix of ponds, canals, and marshes. The wetland habitat supports a dense growth of rushes and sedges that attract hundreds of species. Thronging insects and their larvae make nutritious snacks for fish, bats, and birds. As many as seventy-five times more ducks live in areas with beaver ponds than in areas without them. When beavers vanished, much of this biodiversity vanished with them.

The ecological reverberations rumbled outward. Trumpeter swans lost key nesting sites on top of beaver lodges. Sage grouse had a harder time feeding their young in parched summers. Otters, mink, and moose all had less water to explore. Ponds serving as nurseries for salmon vanished. Basins in the high country lost the fertilizing effects of vegetation impounded by beaver dams. Underground aquifers failed to recharge.

Streams went unfiltered. Floodplains dried out. The consequences were catastrophic. The wetlands that supported up to 80 percent of biodiversity in arid parts of the country literally went up in smoke.

* * *

Alexa Whipple's epiphany about beavers came when pounding out her regular trail run late one spring. Whipple lives in the Methow Valley in Washington state, home of the Methow Beaver Project. The project provides assistance for landowners trying to cope when the world's second largest rodent shows up on their property. This means helping protect trees, installing flow-control devices on beaver dams, and generally doing what's possible to persuade property owners of the benefits of living alongside beavers. When all else fails, the Beaver Project will relocate problem animals. They capture them in traps, house them temporarily, and truck them in pairs to remoter parts of the county.

At the time of her epiphany, Whipple worked in the sustainable agriculture community, showcasing how to grow food in a wildlife-friendly way. The seasoned wildlife advocate talks fast, spilling enthusiasm from every pore. She blows off steam by trail running, a form of exercise increasingly on hold in summer due to wildfires.

In 2014, a series of fires scorched the area where Whipple lives. The Carlton Complex Fire burned a quarter million acres and destroyed more than 350 homes on the eastern slope of the Washington Cascades. Large portions of Bear Creek, Whipple's favorite running route, burned to a crisp. The first time she ran there after the fire, she returned home sobbing. "It felt like walking on the moon," she told me, "and so saddening to think the area would never be what it was in my lifetime." As a wildlife advocate, Whipple saw only a rich habitat turned into ash. "The fire was just so close to home and in a place I know so intimately," she said.

As luck would have it, the Methow Beaver Project had a couple of extra beavers needing a new home. The habitat on Bear Creek was marginal. There were only a few patches of unburnt vegetation remaining. But the Beaver Project needed space at its halfway house, and the biologists thought it worth a try. So they hauled the beavers up to the creek and set them loose.

Whipple was stunned. "The beavers were miraculous over the next two years," she said. "It was so inspiring to see what they could do." Before the fire, the area was hammered by horses and invasive plants. The landscape was dried out and weedy. After the fire, the charred vegetation greened up rapidly. "It came back so beautiful," said Whipple. "There were a lot of seeds still in the soil. They were hiding out under there." Having beavers in the creek prevented moisture and silt from leaking away downstream. It reset the soils to more historic conditions. The dampness jump-started the vegetation. "That moment keeps giving me chills and keeps motivating me," she said.

It also changed the direction of her career. Whipple went back to school for a graduate degree at Eastern Washington University in Cheney and wrote a thesis on how beavers help habitats recover after disturbance. She used Bear Creek as a test plot. On graduation, she scored her dream job and took over as director of the Methow Beaver Project.

When I visited Whipple, she took me up to Bear Creek to show me the recovery. Aspen had regrown into a shimmering carpet of green. Above the aspen waved a canopy of ponderosa pine, their cinnamon trunks charred by the fire. Looking down on the beaver pond from an adjacent slope, Whipple pointed out gray shafts of alders the beavers had felled. The tangled mess looked like an unfinished game of pick-up sticks. The beavers don't care to eat the alders, Whipple told me. But they are savvy enough to understand that felling them creates more light for the tastier aspens and willows to grow.

Whipple led me into the thick of the marsh to see the beaver's work up close. We squelched along the top of an old section of beaver dam before diverting around a tongue of water to get a better view of the lodge. Dense thickets of dogwoods, currants, and Douglas maple made it hard to see where to put your feet. Whipple dodged the beaver canals and the downed trunks with the skill of someone who has learned to think like a beaver. I followed clumsily behind, clutching my notebook and pen and trying not to disappear with a splash.

Up ahead of me Whipple suddenly gasped. "Oh, my god. There is a dam right here. They must be damming up the main channel!" The beaver recolonization Whipple had studied until now had all taken place in a groundwater seep to the east of Bear Creek's main channel. The regrowth was impressive, but what Whipple really hoped to see was beavers moving back into the badly incised main creek. She had just stumbled upon the first evidence of it happening.

Whipple was giddy. "Holy moly. Just wait 'til I show this to Joe" (Whipple's coworker at the Beaver Project). We scrambled along the main channel through thick brush, finding three more dams in various states of construction. At each dam, Whipple shouted with joy. They hadn't been there when she last checked. "That's at least a meter," she exclaimed, admiring the vertical drop created by one dam. "Sorry, but you get to see the reaction of a full-on beaver geek here. This is so exciting." The dams would start repairing the steep banks of the degraded creek by capturing nutrient-laden sediments. An ugly eight-foot gouge would turn into a functioning stream again. "They'll do in six months what would take more than a century to happen otherwise, if it happened at all." It was something the scientist in Whipple termed "low-tech process-based restoration." You don't need human engineers with graduate degrees to restore an eroded creek. If you bring in two beavers, the wildlife will do the work for you. It's cheap, reliable, and effective, and it comes with cute, furry critters thrown in for good measure.

Even for a novice like me, it was obvious how much beavers had improved a place that six years previously had been an ash-covered wasteland. We explored the undergrowth carefully, at one point stumbling upon the skeletal remains of a beaver. The skull's curved orange teeth were still prominent in death. We ran our fingers over the two incisors that had renovated the landscape as completely as any snarling bucket loader. The neatly aligned vertebrae laid out on the ground at our feet had done their share of hauling.

When we finally emerged from the swamp and regained our vantage point on the hill, the light was starting to fade. We looked down on the pond and saw a brown head moving steadily across the surface beneath the felled alders. The beaver climbed out of the water and stood on its hind legs to check out a branch. The well-oiled fur shed water over a pair of smooth haunches. After a couple of sniffs and some exploration with its whiskers, the beaver dropped back in the water and resumed its meander through the gloam.

"Sometimes when I stand here on an evening, I feel like they are checking me out," said Whipple, as the beaver paddled swiftly across the darkening pond. "These are amazing animals," she said, "with a lot to teach."

* * *

Beavers, like wolves, don't appear to be fountains of fertility. They come into estrus only once a year in winter, and the window for conception lasts just a day. If no fertilization occurs, the female can become receptive again two weeks later. If an egg is fertilized, the monogamous pair is well-prepared to settle in for gestation. The adults have plastered the outside of the lodge with damp mud just before the first frost. Its frozen shell will offer a stout defense against predators. They have stashed plenty of aspen, willow, and birch near the lodge's entrance to keep themselves fed. By putting the food pantry underwater, they ensure they won't need to venture out. Safely ensconced in the lodge, they will stay cozy and protected until the female is ready to give birth.

Just over 100 days after fertilization, the mother moves to a special birthing chamber deep within the lodge. Her three to five kits are well-developed at birth. At just under a pound, their eyes and ears are open, and they are covered with waterproof fur. They can swim within half an hour, taking their first dips in water flowing gently inside the lodge. The young venture out after a few days under the careful eyes of their parents. Within two weeks, they stop drinking their mother's milk and start eating solid food.

Attentive parents help any young mammal make it into adulthood. Good siblings are also an advantage. Like wolf pups, beaver kits have the company of last year's young. Sometimes kits from the previous year are around too. Beavers are sociable. Siblings play and wrestle together, honing reactions and motor skills. The older brothers and sisters keep their eyes open for threats, calling in the parents when danger looms. "They really need each other," Whipple told me. They groom each other and help keep the lodge clean. In winter, there can be six or more beavers sharing three square feet of space. Whipple thinks beavers offer valuable lessons in coexistence.

The beaver family stays together until the adolescents start to feel their hormones. They reach sexual maturity at the age of two or three, at which point the young adults set off to find territory of their own, sometimes as much as ten miles away. The matriarch continues to give birth each year. She can produce kits for six or more seasons. Provided they successfully dodge the claws of lynx, coyotes, bears and owls, beavers in the wild can live well into their second decade. Like many rodents, they are reliable reproducers. The moment beaver fur fell out of fashion as a material for felt hats, beavers were ready to surge.

* * *

The caprices of fashion gave beavers one break, and a better appreciation of ecosystems gave them another. Enos Mills, a turn-of-the-century American naturalist, watched beavers closely as they worked the willows around his cabin near Colorado's Estes Park. Mills was struck by the services beavers provided, calling them "the original conservationist." He marveled at how their dams maintained consistent water flow during the summer. He documented how the channels they built created fish habitat. He watched the birds and insects hovering around the edges of beaver ponds. Mills came to appreciate the complex ecological tapestry that beavers maintained. He extolled the beaver's virtues to anyone who would listen. "A live beaver is more valuable to mankind than a dead one," Mills declared. "May his tribe increase!"

Within a couple of decades, several East Coast states had begun reintroduction programs driven by a newfound admiration for the beaver. The programs met with instant success. Beavers were soon a common sight in the Adirondack Mountains, plugging streams and creating wetlands in areas that hadn't seen beavers for a generation. In the 1930s, the

Civilian Conservation Corps led efforts to reintroduce beavers to western states. They released hundreds of beavers in California, Wyoming, Oregon, and Utah. When Idaho Fish and Game wanted to restore beavers to the backcountry, it parachuted them out of airplanes in crates designed to pop open on impact. The western populations, like those on the East Coast, quickly flourished. Freed from the ravages of the market, beavers rebounded from a little over a hundred thousand at the turn of the twentieth century to fifteen million today.

There is no doubt that in many places the absence of predators has made recovery simpler. To a wolf, a beaver is a tasty, calorie-dense snack. An Alberta study found beavers making up a third of a wolf's diet in some areas. One closely watched wolf in Minnesota's Voyageurs National Park ate twenty-eight beavers in a single season. Wolves sometimes den in abandoned beaver lodges, where they might enjoy the lingering musk of a favorite nibble as they take an afternoon nap. A beaver waddling slowly on land is easy prey for bears, coyotes, lynx, fishers, and cougars. The kits have to worry about owls, eagles, and hawks descending from above. As predator numbers grow, beavers will have to become more wary. For now, in many places, their odds remain good.

The competing animals beavers have to worry about most are people. Trappers in some areas still take beavers for their fur, even though the price they receive barely pays gas. A beaver might end up dead in a trap if it gets on the wrong side of a farmer or a highway division. There is no denying they can cause significant damage by blocking a culvert or flooding a field of crops. The U.S. government's Wildlife Services division kills over 25,000 beavers a year. A private cottage industry in nonlethal beaver management is available for those who want to keep them alive. The tide has turned in beavers' favor. Many landowners have seen that killing one beaver simply creates an opportunity for another to move in. Beavers today also have an incalculable advantage they previously lacked. There is an eager army of advocates—known as "beaver believers"—lobbying on their behalf.

In Europe, the North American beaver's cousin (*Castor fiber*) has recovered even more dramatically than across the Atlantic, increasing a thousand-fold. Two dozen European nations have conducted reintroductions, with EU conservation laws pushing things along. Beaver advocacy groups have proliferated, with beavers surfing the same waves of public sentiment as their cousins across the Atlantic. The desire to restore the beaver has reached fever pitch.

One reason beavers curry such favor today is climate change. In a hot new world—one tougher to survive than the Castorocene—beavers help climate-proof the landscape. By delaying the rate at which water flows to the sea, they provide a backstop against drier summers. Land wetted by beavers gives ranchers more forage for their cattle during droughts. Wildland firefighters have valuable fire breaks in valley bottoms greened by beavers. Groundwater recharges better when there is downward pressure from beaver ponds. When land does burn, Whipple's studies found beaver dams slowing the phosphate runoff that follows the inferno.

As the effects of climate change intensify, beavers can literally save lives. Governments are looking for ways to reduce flooding now that five-hundred-year rain events happen every few years. Planting more trees in the high country is one solution. Putting beavers back on rivers and tributaries is another. Researchers in Devon, England, found beavers reduced the flow of water cascading out of the hills during a flood by 30 percent. When a beaver dam pushes high water out of a creek onto a floodplain, it acts as a relief valve does on a pressure cooker. The destructive energy of the deluge dissipates across the fields rather than racing downriver. It is no coincidence the beaver recently became the first mammal officially reintroduced into the UK after a six-year trial. Beaver restoration is not just good wildlife policy. It is good climate policy. Beavers bring with them a long list of benefits. Whipple told me about a bumper sticker common on the Subarus driven by beaver enthusiasts: "Whatever the question, beavers are the answer!"

It is easy to imagine that the beaver's influence on salmon might be negative. Beaver dams certainly look like barriers to streamflow. A dam might block an escape route when a side channel dries up. The water behind a beaver dam can become clogged with silt. This all sounds bad for fish. But Ben Goldfarb, the consummate beaver believer, seriously doubts the overall impact is harmful. "Castorids and salmonids possess millions of years of entwined evolutionary history," he says. "It's ludicrous to me that the harmony of beavers and fish remains up for debate."

Whipple agrees. Salmon are not often troubled by dams. They can go around them, wriggle through, or leap clean across. Beavers, say their defenders, taught salmon to jump. She shared images with me of juvenile chinook and coho salmon swimming within the tangled twigs of a beaver dam. Young salmon are also known to use beaver lodge entrances for shelter, though they must keep an eye open for fish-eating otters that have a penchant for hanging out in the same places.

Other pictures from Whipple's trail camera revealed a menagerie of animals enjoying the habitat created by beaver dams. She showed me images of skunks, otters, raccoons, white-tailed deer, mule deer, mink, weasels, muskrats, coyotes, wild turkeys, mountain lions, and black bears all passing through the rich habitat created by the industrious rodents.

I think the conclusion can be stated boldly. Beavers are more or less required for full recovery of rivers in temperate regions of the Northern Hemisphere. The complexity they create nourishes biodiversity in the system. Fish are one of the most notable beneficiaries. They gain places to rest, feed, and spawn. They gain ponds ringed by vegetation full of the insects and larvae they eat. They gain waters pooling at different depths and moving at different speeds. Empirical studies confirm the benefits. Trout in the UK grew bigger in beaver-modified waterways. Juvenile chinook salmon in the Skagit estuary in Washington occurred at three times the density in pools created by beavers than in free-flowing waters.

A few days after my visit to the Methow Valley, Whipple sent me a photo of two large coho salmon swimming unhurriedly a few inches beneath a beaver. The rodent and the fish looked like a family of three out for an evening stroll. Evolution provided time for these species to work it out. When Tucson's Center for Biological Diversity filed suit to stop the killing of beavers under the Endangered Species Act, it was not the beavers they wanted to save. It was the endangered salmon that needed the beavers to survive. If you want fish back in a degraded river, you could hardly do better than let the river be engineered by beavers.

3 DEINDUSTRIALIZING RIVERS

"When people say Eklutna is in Anchorage, I always want to say, 'Actually, Anchorage is in Eklutna.' . . . Okay!"

The documentary made about the Eklutna Dam removal provided a chance for the Eklutna people to talk back. Tribal Council member Maria Coleman had good reason to be clear about this in the film. Anchorage had long ignored her people. Now was a suitable time for that to change.

The Eklutna Dam came down in 2018. Coleman is proud of the achievement. The Native Corporation's $7.5 million project was a technical tour de force. Engineers lowered heavy equipment three hundred feet into the steep river canyon with a giant crane. Workers descended to the site on a seven-hundred-foot custom-built staircase. Water was diverted through the middle of the work site in giant pipes as there was nowhere else in the narrow canyon for it to go. When the removal was complete, everything was winched back out of the canyon and taken away. Proud as she is of the tribe's engineering skills, Coleman's celebration is still muted. Unlike the dams on the Elwha and Rattlesnake Creek, the Eklutna Dam removal is only the beginning. Getting water back into the river is the next step of the fight.

Eklutna river water long played a role in keeping Anchorage's lights on. The dam was built in 1929 to provide electricity for a city whose population was then only two thousand. From the start, the dam was never going to be enough. As Anchorage's numbers grew, the local utility, Anchorage Light and Power, had to figure out ways to supplement its output. As a way to meet demand, it used dirty diesel generators and, for a time, the engines and back half of an oil tanker wrecked in the 1940s. In 1955, the federal government funded a larger dam at Eklutna Lake together with a 4.5-mile tunnel to carry water to a power plant in the valley next door. The original dam became obsolete. Abandoned and nonfunctional, the Eklutna Dam stopped generating power. But it also stopped fish.

Taking the dam out was a huge victory for those in favor of salmon recovery. But it was only the first of several steps. Anchorage's utility still diverts most of the Eklutna's flow to the neighboring valley. What's left in the Eklutna is just a trickle, too shallow for the salmon waiting at the river's mouth in Knik Arm. When the Elwha dams came out, the river boisterously resumed its natural flow. Big rainstorms pushed it to places it had not been in years. The Eklutna is not so lucky. Thanks to ongoing diversions for hydropower, the fish still don't have enough water. Fortunately, the tribe has an ace up its sleeve. The utilities responsible for hydroelectricity in the watershed signed an agreement with the U.S. Fish and Wildlife Service in 1991 to mitigate their impact on fish. This means, dam or no dam, they have a legal obligation to restore the fishery. The Native Corporation and an environmental group known as the Conservation Fund are determined to see they meet it.

Brad Meiklejohn's work at the Conservation Fund was key to getting the Eklutna Dam removed. Meiklejohn is an athletic-looking outdoorsman based in the Conservation Fund's Alaska office. He is also an avalanche expert and founding president of the American Packrafting Association. It is no surprise Meiklejohn lives and works on the doorstep of some of the most spectacular mountain scenery in North America. Rivers and mountains are his life. When I talked to a friend who is a hard-core kayaking advocate on the East Coast nearly 3,500 miles away, he said, "Oh, yeah. I know Brad." Colleagues laud him for protecting more than 300,000 acres of critical lands. Meiklejohn has the status of senior environmental statesperson in Alaska and speaks with a conviction that forces you to take him seriously. A fisherman friend in Southeast Alaska calls him "the most environmentally ethical person I've ever met." His work on the Eklutna is certainly about salmon, beavers, and river restoration. But it is also about justice.

Meiklejohn describes the Eklutna Dam removal as a project driven as much by ethical considerations as ecology. "The tribe has been beaten down for generations," he told me. "It was time for us to make some recourse." The dam—as well as most of Anchorage—was built on tribal lands, but it never served their interests. While the dam still stood, it was a physical barrier to any talk of fish recovery. The concrete had become a convenient obstacle for those with an interest in dragging their feet. Now that excuse had been taken away. What's left is the water. The trickle that is today's Eklutna River needs to be restored to a torrent.

Although the path to full restoration is now visible, Meiklejohn is frustrated with the pace at which things are moving. "The river's been dewatered for ninety years," he told the local paper. Let's not delay the problem any longer." The hydroelectricity generated by the Eklutna provides only 3 percent of Anchorage's grid capacity. In 2021, the utilities diverted water from their hydroelectric operation down the abandoned river to study what a full flow looked like. If they can restore the Eklutna salmon run, it will help the larger Cook Inlet ecosystem. "Salmon," Meiklejohn says, "don't ask much of us. They only ask that we stay out of their way and don't mess with them. If we do that for them, they keep coming back for us." But when he says "us," he has a particular group in mind. "Eklutna has given a lot to Anchorage's growth over the past hundred years. I think we owe it to the Eklutna people to help restore the river that runs through their community."

The situation on the Eklutna shows that taking out a dam is not always enough. Even throwing in a few beavers won't always cut it. Fish restoration also requires politics—the kind of politics that recognizes fish need clean, free-flowing water and plenty of it. It means a politics committed enough to force a utility to find alternative sources of power, to shut down a fishery for a decade, or to put a few extra cents on a utility bill if that's what it takes. This type of change demands plenty of political will. But it is clear that moving from a system of human engineering and economy back to one of nature's own design can create huge payoffs for wildlife. The animals already know what to do.

* * *

When you move from the Pacific to the Atlantic coast, the names of the fish change, but the stories of crash and recovery remain fundamentally the same. American shad used to run up East Coast rivers in their tens of millions, spawning from January to June in a silver wave of fecundity that washed from Florida to the Bay of Fundy. Shad were a mainstay of the diet of Indigenous people. Quinnipiac, Pequot, and Mohegan Indians built low walls in estuaries to trap schools of fish when the tide went out. Early reports from explorers tell of scooping dozens of shad straight out of the rivers in frying pans. Their Latin name, *Alosa sapidissima*, means "the most delicious herring." Shad and striped bass were once so common that Indians and early European settlers used them to fertilize their crops. During the American War of Independence, the British army blocked shad from running upriver to Valley Forge to starve out the patriots. John McPhee called shad "the fish that fed the nation's founders." Washington and Jefferson sold shad from the Potomac and the Rivanna rivers. Settlers salted them by the hundreds for winter storage.

As European settlement increased, migratory fish on the eastern seaboard lost out to dams, habitat destruction, and overfishing. Massachusetts enacted one of the earliest marine conservation laws in 1649 when it banned farmers from putting bass on their fields. Henry David Thoreau mourned for the declining fish. He sympathized with the shad "still wandering the sea in thy scaly armor to inquire humbly at the mouths of rivers if man has perchance left them free for thee to enter." Shad populations crashed. Thoreau lamented: "Poor shad! where is thy redress?" Smelt, eels, Atlantic sturgeon, alewives, bass, and Atlantic salmon all suffered a similar fate.

After declining for more than a century, state agencies, cities, and volunteer groups have been making efforts to turn the fisheries around.

Some rivers have already been liberated from their dams. The Penobscot in Maine is the East Coast's Elwha. The removal of two Penobscot dams—and the creation of diversion channels and fish lifts on others—has opened up two thousand miles of spawning habitat. Shad are back in their thousands and hundreds of Atlantic salmon spawn again in newly accessible creeks. Alewife numbers have shot up from a few thousand to more than two million, and lamprey are recovering fast. Atlantic sturgeon have been spotted working their way through the bypass channels. On the Kennebec, alewives increased from less than a hundred thousand to more than five million after the Edwards and Fort Halifax dams were taken down. Striped bass, shad, and American eels are back on the river together with the bald eagles and osprey that prey on anything too close to the surface. A Gulf of Maine recovery plan for endangered Atlantic salmon released in 2019 will continue the dam removals and make culvert upgrades on Maine's creeks to improve spawning access.

Further south, dams have fallen to jack hammers, excavators, and dynamite at a rapid rate. Pennsylvania alone has removed nearly 350 dams. But as you move south and the land use becomes increasingly intensive, the work of recovery becomes more complicated. Most East Coast rivers flow past farms, cities, and industrial sites. Millions of people live within their watersheds. Even after dams are removed, rivers must be cleaned and riparian habitat restored if fish are to recover. It is difficult and time-consuming, but the payoff can be high.

Reductions in runoff from farms and cities bordering the Chesapeake Bay have helped seagrass beds nearly triple in size over three decades. The seagrass acts as a nursery for young fish as well as being an effective carbon sink. Aided by harvest restrictions, striped bass have recovered well enough to restart several shuttered fisheries. Juvenile American shad in the Potomac climbed to fifty times their low point in the 1980s before leveling off. The Rappahannock has also exceeded its shad recovery target.

Not all the news about shad is encouraging, however. In the Delaware and the James, recovery remains tenuous. Even in the Potomac, there are concerns about shad's long-term viability due to climate change, invasive species, and overharvest. A 2020 report by the Atlantic States Marine Fisheries Commission on the status of shad concluded the species remains depleted. Many people think more dams need to come down for their prospects to improve. Unblocking fish passage and removing pollution so spawning areas can recover is essential if fish are to stand a chance. But aggressive clean-up and streambank restoration can work. In Europe, the Seine, the Thames, the Mersey, and the Rhine all have salmon back in rivers previously too fetid to support them. And if you still need evidence of the benefits river restoration can provide, a visit to the banks of the Potomac in August might just provide it for you.

* * *

The Potomac has been called "the nation's river." George Washington's home at Mount Vernon, the Jefferson and Lincoln memorials, the Watergate Apartments, and the Old Stone House all stand on its banks. The river used to be an ecological haven. An early English visitor called it "an Edenic waterway teeming with wildlife." By 1965, Lyndon B. Johnson described it as a national disgrace. When the Clean Water Act passed in 1972, no one dared dip a toe in its filthy waters. Wildlife had either been killed off or stayed as far away from the Potomac as possible.

After half a century of restoration, the river has turned a corner. Several dams on the Potomac are gone, governments regulations have reduced pollution and slowed agricultural runoff, and volunteers have restored streamside vegetation. Wildlife has responded. Bald eagles now breed within sight of the Washington Monument. Blue crabs, beavers, seahorses, and ravens have all returned to the Potomac. Fishermen can catch American shad, striped bass, white perch, and smallmouth bass. The most startling returnee of all, however, is not a fish. It's the bottlenose dolphin.

Over the past decade, the dolphin population in the Chesapeake Bay has grown dramatically. The Potomac-Chesapeake Dolphin Project photographs the fins of dolphins and matches distinctive marks and notches to databases from other parts of the Atlantic seaboard. The project names the dolphins after political figures. There is a Nancy Pelosi, a George Washington, and a Barbara Bush. Project staff identified two hundred dolphins at the Potomac's mouth when they began counting in 2015. Three years later, the number had risen to a thousand. Dolphins now nose upstream to within fifty miles of Washington, D.C., something that has not occurred since they reached Alexandria in Washington's time. In 2019, a researcher saw a dolphin give birth inside the river. It was only the third dolphin birth ever witnessed in the wild.

Nobody is quite sure how to explain the rebound. The Marine Mammal Protection Act prohibits harming dolphins in U.S. waters and has certainly played a part. Pollution in the Chesapeake Bay has also been reduced. Warmer ocean temperatures might be moving dolphins north from the Carolinas. Sea-level rise in the Chesapeake Bay is twice the global average. Heavier rainfall is putting more fresh water into its rivers. The dolphins have cleaner water and more to eat. Whatever the reason, dolphins heading upriver toward the American capital is a powerful symbol of recovery. When Lyndon Johnson was president, no one would have dreamed of seeing dolphins in the Potomac. The change has been lightning fast. "Dolphins are swimming in the Potomac," declared one recent headline. "Are we next?"

* * *

The lessons learned from successful river recoveries are clear. Keep your interventions small. Maintain as much complexity in the system as you can. Protect, preserve, and restore streamside habitat. Mimic how the original system worked. And show some humility! In ecological systems as complex as rivers, humans bump up against the limits of their management skills. Our species is good at engineering lots of things. But rivers don't seem to be one of them. It may be time to admit this. A better way to show humility in the face of river management is to ask yourself, "WWBD?" What would beavers do?

Andrew Lahr has taken the idea of learning from beavers to heart. On a September morning thick with dew, I joined the twenty-seven-year-old New Jersey native for a morning on Montana's Lolo National Forest gathering data for his Ph.D. research on creek restoration. The method

Lahr and his U.S. Forest Service partners had chosen for their restoration was nothing if not humble. They were trying to mimic a beaver.

Lahr is lanky and bearded. He has the kind of physique that, even when he's wearing stout rubber boots, allows him to jump ditches and barbed wire fences that leave the rest of us stranded. I was already dubious about the chances of keeping my feet dry, given beaver mimicry was the order of the day. My doubts increased when I saw Abigail, a U.S. Forest Service Americorps volunteer working with Lahr, pulling on chest waders.

The study site was on a degraded tributary of Howard Creek five miles up a bumpy gravel road near the Montana-Idaho border. The creek wound through an open meadow surrounded by high-angled slopes of Douglas fir, Englemann spruce, and lodgepole pine. Nobody was quite sure how the creek became so degraded. Up here, pastures got cleared and livestock were grazed for decades without anyone writing down the details. The consequences of the grazing were less ambiguous. Halfway down the meadow, the alders and willows disappeared, the vegetation simplified into a carpet of sedge, and the narrow creek disappeared into a deeply cut channel. "From here on down, this place is a disaster," Lahr told me when we walked into the top end of the site.

With the help of some burly workers from the Montana Conservation Corps, Lahr had altered the flow of the creek with a sequence of thirteen beaver dam analogues (BDAs). At carefully selected sites, the volunteers had pounded a couple of dozen stakes perpendicular to the current in two rows a foot or so apart. They wove branches and twigs between the rows to provide a scruffy barrier to the current. Over time, the tangle of branches caught more and more debris floating down the creek, and, before long, the structure became a pretty good surrogate for a beaver dam. "We are taking a stream that is degraded and messing with it a little bit," Lahr told me. "We look at what beavers are doing and ask how we can do that too." It was a homespun version of what the beavers in the Methow Valley were doing at Bear Creek.

I was here to watch Lahr and Abigail take their monthly measurements of how the BDAs were doing. All around the creek, the meadow grasses had taken on the burnished copper color of fall. It was the time of year when the morning air had a chill that you knew would be gone by lunchtime. Accompanying us on the outing was Lahr's chocolate lab Barley, named—for reasons Lahr was rather vague about—after the primary ingredient in beer. Quickly bored by the data gathering, Barley kept up his own version of beaver mimicry for most of the morning, chewing enthusiastically on a stick at every BDA we examined.

Lahr was tracking what the BDAs did to water flow through the meadow. This meant taking measurements at the surface as well as beneath the ground. Spaced across the meadow were lines of five or six PVC pipes pounded into the soil in transects. Abigail and Lahr methodically dropped a sounder down each tube until it beeped, indicating where the groundwater began. Then they jumped into the creek by each BDA with a marked wooden pole and took seven or eight measurements. They recorded beaver-relevant variables like downstream plunge, wetted width, and upstream pool depth. This was the time when everyone tried not to get a boot full of water.

Lahr also wanted to know the impact of BDAs on fish. Another technician funded by his grant was wading around with an antenna somewhere nearby, trying to detect the six hundred trout Lahr had equipped with passive integrated transponders—what fisheries technicians call PIT tags. These transponders give off a unique identifying code that can be picked up by an antenna. By noting where in the creek each fish was swimming, Lahr could track how well different species of trout were moving through the dams. All the data went into an iPad to be analyzed back home in the lab.

Gathering this much data takes considerable time and effort. Six different sites and dozens of BDAs needed to be hit multiple times over the summer. I admired Lahr and Abigail's diligence and asked Lahr what kept him motivated. The consummate scientist made it clear that if you are going to "rebeaver" the west for ecological reasons, you'd better have the data to back up your hunches. "As much as I'm a beaver believer, I try to be level-headed. I am a scientist, and I can't just go around spouting the good word all the time. I want to make sure we are doing it right." Other researchers partnering on the study were measuring the carbon stored in the rewetted meadow and the methane it emitted. One uncertainty about this new restoration method is whether beaver dams could increase natural methane emissions more than they contributed to carbon storage. Soggy and decomposing vegetation is renowned for belching climate-warming gases.

When I asked Lahr why he was building BDAs rather than reintroducing real beavers, he said part of the reason was habitat. Creeks eroded too far below ground level no longer spilled onto their floodplains. As the floodplains dried out, the aspens and willows favored by beavers had disappeared. "BDAs," he told me, "can jumpstart a dry meadow and reconnect a stream with its floodplain. A beaver might then establish itself on its own, or you could plop a beaver there once the habitat is suitable." Politics also played a role. Some state wildlife agencies are less enthusiastic about spreading beavers across the landscape than others. They worry about conflicts with landowners. Creating analogues can be a less controversial way to reengineer a river.

Seeing the effort that went into the restoration of this one tributary of a single creek in a massive drainage made it clear what a gargantuan task lies ahead to make river systems whole again. There are also plenty of wild cards that could derail fish recovery every step of the way. Hotter, drier summers raise water temperatures too high in some rivers for bull trout and salmon. Invasive species such as brook trout create competition for struggling natives. When fish head to the ocean, changing food availability caused by warming temperatures disrupts their growth patterns. Nutrients no longer cycle through the marine ecosystem the way they used to. Food sources in some places have disappeared.

The magnitude of the task ahead is one reason to look for natural allies like beavers. Ben Goldfarb, unabashed about his beaver cheerleading, lives by the slogan "Let the rodent do the work." If beavers were restored to one-tenth of their numbers before fur trapping began, this would create an army of forty million highly skilled engineers working on creek restoration across North America for free. What's more, beavers are the best there is. "Beavers know about stream health and are constantly restoring," Abigail remarked as she and Lahr worked through the transects, punching numbers into their iPad. It's what they do for a living. "If we increase the abundance of any native engineer," she said, "it's a good thing."

Good habits for living alongside rivers have largely been forgotten in most settler cultures. Though not, Abigail pointed out, by Indigenous people. As part of her training on river recovery, she had visited tribal restorationists on the Blackfeet Reservation. The Blackfeet are known for having refused to help the Hudson Bay Company trap beavers in the eighteenth and nineteenth centuries. Beavers are one of the three original animals in the Blackfeet's creation story. Their beaver ceremonial bundle, carried by tribal elders to channel the beaver's power and guide their prayers, is one of the most complex bundles in North America. The Blackfeet know how important beavers are to the landscape. Today the Ksik Stakii project on the reservation is helping restore the beaver as a tool to fight climate change. High-schoolers are putting on hip waders and building beaver dam analogues to rewet the reservation's portion of the Rocky Mountain Front. River restoration, like on the Elwha and Eklutna, is also cultural restoration.

The question for rivers and the species that depend on them is how far this type of relearning can go. Even in the age of dam removal, elements of the engineering mindset are tough to shake. While working hard to recover a river naturally with one hand, our species seems determined to retain an industrial grip with the other. And nowhere is this mix of the natural and the industrial more evident than in fish hatcheries. Even as we let go of dams, we seem reluctant to let go of hatcheries. This brings the story of river restoration back to the Alaskan salmon encountered earlier, which was driven by hormones to ascend a cement channel in a Prince William Sound hatchery, its bulging flanks protecting a belly full of eggs.

* * *

I forged a bond with Toby, my skipper, right away. He was a shrewd businessman, a deep thinker, and an eagle-eyed fisherman. He stuck out in the crusty, hypermasculine world of commercial salmon seining.

A small, delicately featured Californian, Toby ran his boat a little differently. We had an espresso machine on board, immaculate southwestern-styled seat covers, and a high-end stereo. Thoreau's *Walden* and Terry Tempest Williams's *Refuge* were part of the boat's library. His crews were mixed gender. Unlike most boat skippers, he never yelled at any of them. Toby owned a yellow lab named Sweetie who sometimes accompanied us on board, often wearing a pink collar.

Although his style rubbed some captains the wrong way, when Toby scouted for salmon from the crow's nest of his forty-seven-foot seine boat, the turbo-diesel engine sang. Some boat captains choose a preferred spot where they grind out their day laying the net methodically on a repeat cycle. Not Toby. He would race around the bays hunting for the big schools of fish. With his expensive polarized glasses scanning the surface, there was no one in the fleet who could more reliably encircle a wandering mass of salmon than Toby.

For several summers, I crewed on Toby's boat doing cost recovery at salmon hatcheries in Prince William Sound. Although the bulk of the fish making their way back to an Alaskan hatchery are available to commercial fishers, a portion are reserved for a boat contracted by the hatchery. The hatchery seiner is given first dibs at the harvest by the Alaska Department of Fish and Game. It fishes closer to shore than other vessels, scooping up fish and selling them directly to a delivery tender camped out nearby. The tender sucks the fish straight out of the seiner's net with a hose lowered into the water like a giant elephant's trunk. The sale of these fish—together with a small tax paid by the fishermen—allows the hatchery to recoup its expenses. The seiner gets a decent cut while avoiding the free-for-all of the commercial boats farther out in the Sound. The hatchery also keeps a portion of the fish alive for breeding stock. For a few years, Toby had a lock on several hatchery contracts.

Seining salmon close to a hatchery is a delicate art. The water can be shallow. Rocks are everywhere. The skipper needs to be vigilant. Hungry sea lions pop in and out of the net as you draw the bag tight. It's not unheard of to encircle a whale. The sheer volume of fish schooling up in the bays creates a dangerous weight. The hydraulic lifting equipment on board generates potentially lethal forces. The captain needs to make good judgment calls to avoid expensive and dangerous mishaps.

Every seine boat travels with a skiff—a smaller open boat—in tow. For a couple of seasons, I was Toby's skiffman. The skiff helps lay out the thousand-foot-long seine. It keeps one end taut while the bigger boat fashions the rest of the net into a U-shape to gather the salmon. When fishing near land, the skiff driver often places the nose of their boat against a rock while the larger vessel tows the other end a couple of hundred yards offshore. Fish moving down the shoreline suddenly find themselves swimming into the curve of a net. At the skipper's signal, the big boat and the skiff come together to trap them in a circle of webbing. When the loop is complete, the skiff hands off the end of the net and heads to the opposite side of the big boat, where it clips into a harness. By pulling the boat gently away from the circle of seine net, the skiff counteracts the big boat's tendency to get entangled as it starts to haul the heavy webbing aboard. Over a season, this mixture of physics, art, and instinct becomes a well-practiced routine.

As skiff driver, I would watch from my position on the end of the harness as the deck crew hauled the net on board and the circle left in the water became smaller. The salmon always took a while to recognize their plight. When the purse seine got tight enough, the seething mass of salmon started to twitch. Fish on one edge of the net would panic and break the surface. Like a murmuration of starlings turning simultaneously in flight, a current ran across the net until the whole surface exploded with thrashing fish. Frightened salmon rushed en masse to one side, forcing down the corks responsible for keeping the top of the purse afloat. The sunken portion became a relief valve that drew further waves of fleeing fish. As the escapees poured over, more corks disappeared, and the purse started to collapse. This was money that needed to be recaught and gear that risked being broken.

When this happened, Toby would stop hauling the net and gesture for me to help. After detaching from the harness and gunning it quickly to the sunken part of the net, I would put the skiff in neutral and reach for the submerged corks with my arm or a boat hook. I'd haul the line back above the surface and lash it to the gunwale of the skiff. The escaping fish bumped against my arm as I pulled. In the background, Toby would flash an anxious smile as the liquid world of the fish ran momentarily into our own. Even twenty years later, I remember the quality of those moments. Boundaries evaporated. The sculpted bodies of marine life and the brute power of winches briefly collided. Everything blended into a singular mass of life and struggle. Person and place fused together forever.

Hatchery fishing was a way to make good money outdoors and forge lasting friendships. It put me in closer contact with wildlife than I had ever been before. The profusion of marine life I encountered in Prince William Sound seemed to display the world as it was meant to be. At the time, I saw no dark clouds.

Two decades on, I'm not so sure. The removal of the Elwha Dam was nearly derailed when environmental groups filed suit at the last minute to prevent the Lower Elwha Klallam from putting hatchery fish from a different river system into the Elwha. The tribe ended up changing course and stuck with local fish. The idea that a hatchery should participate in the Elwha recovery at all created controversy from the start. When you want a natural system to repair itself, perhaps you should refrain from doing it with industrial methods. Like the tide against dams, the tide against hatcheries has been turning.

The key to the success of wild salmon is their genetic diversity. Pacific salmon have only one chance to spawn. When two wild salmon take different forks of a river, they are sealing their progeny's fate. Fish can't hop out of a creek and walk over a ridge into the next drainage. Two fish in different branches of a river might as well be swimming on different planets. The genetic differences that accrue in subsequent generations are much greater than the differences found between people from different towns and countries.

Over the generations, salmon become fine-tuned to the conditions of their native creek. Each creek has a slightly different temperature, a distinctive food supply, and a unique timing for spring snowmelt. Rocks polished by the descending waters give the creek its own scent.

Fish adapted to different drainages ensure abundant genetic diversity across a region.

This diversity is why salmon managed to sustain Indigenous people for so long. Different runs are suited to many different conditions. When one run fails or is hit by a disease, another booms. The population as a whole can ride out the bumps. The tiny percentage of returning salmon that stray from their native creek ensure enough genetic mixing to keep the salmon strong. The prolific reproduction cycle—millions of females beating out redds on creek bottoms, billions of eggs bathed by the cold milt of the males—keeps the genetic engine turning.

Hatcheries compromise this diversity. Salmon eggs incubated in a hatchery have a survival rate ten or twenty times higher than eggs wintering on the bottom of a creek. But when it comes to fish returning from the ocean several years later, the odds are reversed. Wild fish have the smarts to survive in the ocean. Hatchery fish are less finely tuned. There is little selection pressure in a hatchery to promote survival in a wild environment. Increasing evidence suggests hatcheries slowly degrade the genetic health of fish resulting in lower reproductive success at the end of their lifecycle.

As the genetic strength and resourcefulness of the salmon decrease, their long-term prospects deteriorate. This creates knock-on effects in the system. Hatchery fish tend to be smaller than wild fish. Killer whales and other predators have to work harder for the same calories. The remaining wild fish have more competition for food, and their genes are unnecessarily diluted when they interbreed with their industrially raised cousins.

One hatchery manager I met in the lower forty-eight spoke to me about his increasing doubts. He spent part of his career in Prince William Sound before moving south to work with endangered fish runs on the Columbia. "I'm not sure I agree with programs that put hundreds of millions of hatchery fish into the ecosystem," he said. Swamping an ecosystem with inferior fish for commercial gain didn't seem to him like it would end well. Even doing it as a temporary bridge after dam removal raises questions, like those asked on the Elwha. The river knows how to create healthy fish better than we do—slowly and steadily. A *Los Angeles Times* opinion piece on salmon in the Klamath put it like this: "Allowing hatchery salmon to mix with struggling native salmon after removing the dams is like rescuing a dying man only to slowly poison him."

Fisheries experts have spirited debates about the value of hatcheries in situations where a system is under intense stress. The Lower Klallam Elwha Hatchery might have been necessary to ride out the sediment pulse when the dams first came out. In the Columbia system, some runs would almost certainly not exist at all were it not for the hatcheries. The right answer probably varies by location. Alaskan hatcheries might be able to reduce their impact on wild fish through smart siting, careful timing of releases, and vigilant sourcing of their brood stock. On river systems where salmon stocks have been eliminated, a hatchery might be able to restart a run. But questions linger. A hatchery manager in Southeast Alaska told me there is a huge need for research. She previously worked at a hatchery whose incubation trays are now so prone to disease they have to take the eggs elsewhere to hatch. With climate change altering precipitation in the mountains, the likelihood of encountering these kinds of problems is increasing.

One thing seems certain. Five billion hatchery fish released in the North Pacific each year between California and Japan provide significant competition for their wild cousins, whether or not they have a genetic cost. Some fisheries managers have already seen enough. Natural Resources Wales closed all its salmon hatcheries in 2015 for lack of evidence they helped. The investigative documentary *Artifishial* takes a full-on swipe at hatcheries, claiming they do more harm than good. The film took its subtitle from a twenty-five-year study on the effect of hatcheries on Snake River salmon—*The Road to Extinction Is Paved with Good Intentions*.

* * *

The debates around dams, hatcheries, beavers, and river recoveries can suck you into an endless whirlpool of uncertainty. The science is complex. The politics is gnarly. The "what ifs" are numerous. It is easy to despair about the mismatch between political and evolutionary timescales. But looking at the debates highlights at least one important philosophical point. It showcases the contrast between an engineering approach to a river system and a natural approach. It should come as no surprise that a dynamic, intricately branched liquid habitat full of genetically distinct forms of life is very easy to mess up. Dams and hatcheries date from the industrial revolution, bringing the same mindset to the management of water and the production of fish as to the production of steel beams and railroad carriages. The idea you can compensate for a blocked river simply by producing replacement fish in a hatchery risks putting an industrial glove onto a biological hand. It's a mismatch worth pondering if other wildlife recoveries are to get off the ground.

Back in Eastern Washington, the Methow Beaver Project still provides the odd beaver with a temporary home before relocating it to a stream nearby. Alexa Whipple is keen for the organization to switch its emphasis from *relocating* beavers to figuring out how to live with them. She knows the long-term success of beaver recovery depends on coexisting with beavers rather than moving them far away. This requires spending lots of time on the property of irritated landowners listening to their gripes. Whipple then works out how to MacGyver various flow devices and PVC pipes to stop valuable lands from being flooded. The work requires careful listening and plenty of goodwill, traits Whipple brings to her work in spades.

The beavers in the Methow wait their turn for relocation in an unlikely setting. They swim up and down the cement raceways of the Winthrop National Fish Hatchery. The hatchery happened to have ponds available with a good water supply when the beaver project started. The raceways are long, straight rectangles, and the beaver "lodges" are made out of breeze blocks. The hatchery still releases hundreds of thousands of young salmon and steelhead into the Methow River each year. It does this to compensate for the dams lower down on the Columbia River. It's a stopgap—but, many think, necessary—measure. The hatchery manager told me that without the artificial boost from the hatchery, the natural fish runs would likely peter out.

For now, it creates a bizarre juxtaposition that captures perfectly the complexity of river restorations. Not far from where salmon smolts are growing, beavers swim up and down in cement channels, the lucky ones accompanied by a new mate. When they dive, their thick fur streams bubbles as they wait in their quadrangular home for an opportunity to turn impoverished creeks into healthy ecosystems again. A couple of channels over, artificially hatched salmon are feeling the itch to start heading downstream through a gauntlet of dams. They mill slowly around in their thousands, sucking down pelletized food, unaware of how close they are swimming to some of the planet's most skilled river engineers.

FOREST MANAGERS

1 AN OWL WITH A PARACHUTE

I could hardly believe my luck. After coming down in sheets for most of the day, the rain stopped and the clouds lifted just in time to create the landscape of my dreams. The late-September understory in Washington's Okanagan-Wenatchee National Forest was a pixilated yellow and rust. The conifers were refreshed after the day's soaking, raindrops still clinging to hemlock needles and wolf lichen. It was cold enough for the peaks to wear an early mantle of snow. Their crests glowed tangerine in the evening light. The kaleidoscope on display sucked me in completely. But I had other things to focus on. I pulled my eyes back to Highway 903 just in time to see the tail lights of the U.S. Geological Survey truck ahead of me turn off the pavement and head into the forest.

When I stopped on the side of the dirt road ten minutes later, Melissa Hunt was already out of her truck and setting up the equipment. "Barred owls like drainages," Hunt said, nodding in the direction of the gully that fell away from the road a few feet from where we parked. In the fading light, she placed a speaker loaded with barred owl calls on the roof of her rig. Hunt entered the time and weather conditions on her data sheet and activated her hand-held transmitter to start the fifteen-minute cycle. The speaker began shrieking a sequence of calls designed to make nearby owls think there was an intruder in their territory. Hoots the biologists named "pair duet," "eight note," and "banshee" pierced the thick woods. Hunt occasionally adjusted the direction of the speaker to cover the most ground. Our job for the night was to document which of the different hexagons marked on Hunt's GPS contained barred owls.

"We will be in areas new to me tonight," said Hunt. "For most of this study, I was a remover, not a surveyor."

I tried hard not to blink.

"Remover" was an accurate term for what Hunt did. But it was a euphemism. Hunt is one of the best in the business at shooting barred owls out of trees with a shotgun. The twenty-eight-year-old, slightly-built wildlife management specialist from Belmont, New York, had spent five winters tracking barred owls and systematically blasting them from the canopy with a twelve-gauge. The goal was to reduce the barred owl population enough to relieve the pressure on spotted owls. It was a divisive study generating high emotions on all sides. But Hunt loved the work. "I'm kinda sad the removal part is over," she said.

I wanted to meet Hunt because her work prompts a difficult question hovering over a range of wildlife recoveries. How much manipulation of a system is permissible to help a species return? I assumed wildlife must survive on their own to truly count as wild. But we live in an age when many cannot. In forest environments on both sides of the Atlantic, I found owls, bears, wildcats, and vultures all highly dependent on an array of interventions. This conflicted with a strong intuition I held about the need for wildlife to live independently of us. But I was starting to wonder if the intuition was wrong. And the owl prompting Hunt to wield her shotgun was charismatic enough that it might just help change my mind.

* * *

The spotted owl is one of thirteen hundred species listed under the U.S. Endangered Species Act. Its five subspecies range from British Columbia to Central Mexico. Northern spotted owls (*Strix occidentalis caurina*) are chestnut-colored with a generous spackling of white spots. The medium-sized predator has a wingspan up to four feet. Spotted owls have the sharply hooked beak typical of owls and dark eyes set symmetrically on a prominent facial disk. They live mainly in coastal forests and are highly dependent on the cavities and broken tops of old-growth trees for successful nesting. They do most of their eating at night by perching on a low branch and using their sharp eyesight to detect flying squirrels, voles, and woodrats. The owl swoops silently from its vantage point and grabs the prey in its talons, a strategy known as "perch and pounce." Northern spotted owls hate any disturbance of their forest home, something they are finding increasingly hard to avoid.

In the 1990s, the dwindling owl population was the focus of a heated debate about how to manage the northwest's remaining old growth. Spotted owls needed the giant spruce and fir to nest. Loggers wanted the trees to prop up a struggling industry. After a testy summit attended by President Bill Clinton and Vice President Al Gore, a new Forest Service policy reduced logging in the region by 80 percent. Logging companies packed their bags and moved to the southeast. It left a bitter taste in many mouths and turned the spotted owl into a hero or villain, depending on where on the environmental spectrum you stood.

Two decades later, despite the reduction in logging, the northern spotted owl is again in trouble. The biggest threat to their survival this time is not a logger with a chainsaw. It is a bigger, more aggressive owl from America's East Coast that has moved into their territory. After the political battles of the 1990s, this turn of events has been a cruel blow to all involved, not least the spotted owls themselves. Biologists have wondered whether there is anything they can do about it. Hunt is at the epicenter of a highly controversial experiment in wildlife management to find out.

Whatever you thought about the experiment, Hunt had the skills for the job. Her dad taught her to shoot when she was seven. He instilled in her how to appreciate the woods as a hunter and an outdoorswoman. There was no question she was going to study wildlife in college. Hunt graduated from the State University of New York at Cobleskill with a wildlife management degree and was hired by a contractor who specialized in the control of problem animals. She was assigned four regional airports in New York and New Jersey together with a nearby city landfill.

She kept gulls, deer, foxes, and woodchucks off the runways. She caught snowy owls at Buffalo-Niagara International Airport and released them a safe distance away. She used bangers and traps to help out with the landfill's crow problem.

"My time was filled with lasers, paintballs, and pyrotechnics," she told me with a smidgeon of glee. She also tried using drones to scare persistent offenders away. The wildlife were smart and quickly learned when she was bluffing. If all else failed, she stepped in with "lethal techniques" to persuade the remaining animals they needed to take her seriously.

Taking her skills to the owl project was a natural next step. She was the first female remover and the youngest on the team. Snowshoeing solo in winter through the northern Washington forest is not for the faint of heart. But Hunt is no-nonsense and is not intimidated by the woods. She had killed her first bull elk from eight yards with a bow a couple of weeks before we met and spent nineteen hours packing it out of the Idaho backcountry with her boyfriend. "I like being outside," she said with a shrug.

The removal part of the experiment had just finished, and now the researchers were making their way through the control areas to see how many owls lived in untreated parts of the forest. For Hunt and her coworkers, this involved night after night navigating rough forest roads and trails to survey each hexagon marked on the map. She spent long evenings with only the trees and wildlife for company. Hunt had seen tons of deer and elk, a bear or two, and even a mountain beaver—a critter that resembles a marmot more than it does a beaver. One night, a cougar ran alongside her truck for several seconds before bolting back into the woods.

She knew the project was contentious. When people asked, she usually told them she worked in wildlife research. Her friends understood the need for the study. But the terms "owl-Qaeda" and "owl terrorist" had been thrown her way and not always in jest. Shooting owls was controversial. "You never know who is going to get upset," she said.

When I asked her how she felt about the barred owls she was removing, she replied, "I like them. . . . I just don't like them here." She had formed an attachment to the charismatic raptor. "I have a strong appreciation for barred owls and have spent a lot of time with them. They command a lot of respect. They are no pushovers."

One particular owl, she told me, had avoided her assiduously. "He went through three years of me removing every mate he had." She saw him once in the first year, but after that he never came close enough to get a shot. She felt for him. But the ecologist in Hunt saw no option. Spotted owls were smaller and more timid and needed three times the home range of barred owls. The barred owls harassed them and stole the best nesting sites. On rare occasions, they killed them. The northern spotted owl population was plummeting. One owl biologist described the species as "circling the drain." Barred owls, in stark contrast, were exploding. Hunt felt an obligation to act. "It's a tough pill for many to swallow," she conceded. But the alternative was to give up on spotted owls, something Hunt dismissed as the easy way out. Every time she pulled the trigger, she felt remorse. But her commitment to the spotted owl's survival kept her going.

It was a difficult management dilemma, one that brought the human role in the survival of wild animals into focus. A vulnerable species needed a hand if it were to stand any chance of recovering from a precarious position. But the helping hand was not benign. It was wrapped around the stock of a well-oiled firearm.

* * *

I reached the lead investigator in the study, Dave Wiens, by phone in Oregon to talk through the ethics. He shared Hunt's dedication to the cause.

"It's nothing that's enjoyable to anyone—going out and shooting owls. That's for sure. But from what I have seen, they have major ecosystem impacts." Barred owls may be cute, but they are ruinous in the wrong environment. "They are apex predators," said Wiens, "and they are new to the system." As formidable opportunists, Wiens told me, they had tapped into a niche that had not been fully tapped before. "They exploit aquatic prey species: amphibians, fish, snails. There are not many nocturnal predators that have exploited that particular environment. Many of these species are naive to nocturnal predation all together. And this isn't to mention the prey species they really focus upon—like the small mammals and flying squirrels that are also important to native predators in the Pacific Northwest. They eat basically anything that moves in the forest within a particular size range." And barred owls were booming. Really booming.

The picture Wiens was painting didn't match my basic understanding of owls. Owls have a celebrated place in children's stories and folklore. They sit patiently on snags and hoot in front of a full moon. They revolve their heads to see what is going on behind them, while wearing monocles and dispensing wisdom. They usually perch alone. They don't invade old growth and certainly don't wreak havoc on forest ecology. I had never thought of them as ruthless predators with devastating effects on native species.

The reality was something different. Barred owls had spread like a blight across the Pacific Northwest over the last half century. There were now more than 3.5 million of them across the country, making them one of the most numerous U.S. owls. Some stretches of old-growth forest were coated with a feathered blanket of barred owls.

"In the Oregon Coast Range, we see incredible densities of birds," Wiens said. "They are so thick. No matter where you go on the Coast Range, if you are in the forest, you are going to be standing in barred owl territory." At this density, they become a devastating aerial army. "During nesting season, they will have family groups of five or six birds per territory, and you have thousands and thousands of these territories spread across the landscape. It's not hard to envision the impacts they must be having by clearing out the forest of prey species. It triggers in my mind a whole cascade as they deplete the resources." Listening to Wiens turned my primitive understanding of the owl dilemma on its head. The need to manage barred owls had very little to do with the political capital invested in the spotted owl. It was because barred owls were wreaking havoc on the food chain. This proliferating species needed serious management if spotted owls were to survive.

* * *

The forests around Cle Elum where Hunt operated did not have the barred owl density of Oregon's Coast Range. "We have to work harder for our owls," she said with a dash of pride. After five stations without any sign, I wondered if today's rain meant we were going to get skunked. Owls don't fly much when it is wet. They don't like the noise made by their damp feathers. At the sixth station, just when Hunt was getting to the crux of the story about her recent elk hunt, the recording on the speaker was suddenly interrupted by a more urgent call coming from somewhere to my left. Barred owl! Hunt switched on her powerful flashlight and flicked it around the nearby trees. We craned our necks upward. She stopped it on a snag right next to the road.

"There she is," said Hunt.

I followed the beam to the tree and saw . . . absolutely nothing. The gray wood of the fifty-foot trunk came to an end with a couple of short stobs and what looked like a rounded husk of bark.

"Where?"

"Right in the beam. Plain as day. You can see the shine in her eyes."

I still saw nothing. I always thought of my eyesight as decent, and the longer I looked, the more embarrassed I started to feel. Then the husk leant to one side for two seconds before moving back upright. I grabbed my binoculars and scanned them up the trunk of the tree until I reached the top. The husk had transformed into an owl, its chest feathers dappled with streaks of gray. Through the binoculars I could see a pair of black eyes glistening in the light.

"That would be a near perfect shot," said Hunt. "Twenty meters or less. No branches blocking the way."

It was a thrill to see the owl up close. As the recorded calls kept playing from the top of the truck, the owl leaned from one side to the other, calling back, trying to work out what was going on. She wanted to know who to challenge. If there was a pair, Hunt told me, the female almost always came in first. (Hunt muttered something about "wimpy males.") The females were noticeably bigger than the males and, as with most raptors, bolder and more aggressive. Because owls fly so quietly, the females often deliberately smack their wings into a branch when they land—Hunt called it a limb crash—to broadcast to the intruder that trouble had arrived. More than once, a female had struck Hunt's caller with its talons. Other surveyors reported hats knocked off their heads by angry owls protecting their territory. These owls can be ornery. There is even a case in North Carolina where some experts are convinced a barred owl caused the death of a woman found in a pool of blood with frightening skull lacerations.

The trick when planning to shoot them, Hunt told me, was to keep the owls interested until they sat on a tree long enough for her to get a good shot. This meant varying the calls on the speaker, something she could do with her hand-held transmitter. The owls moved around trying to get a better view of what was going on. There was a call known as the "goodbye hoot" that meant an owl had seen enough and was about to leave. If possible, when a pair was involved, Hunt tried to shoot the male first. The female would then swoop in close, "blind with rage," as Hunt put it, offering her a chance to get the pair. If she shot the female first, the male tended to keep its distance. I took Hunt's word for it. She told me she had removed over 350 owls in the course of her work.

Three stations later, the dynamic played out exactly as Hunt had described. With the caller shrieking on top of the truck, a pair of owls came in to investigate the potential intruder, with the female leading the way. The female's calling was noticeably more aggressive. She kept moving from tree to tree in the glade where we stood. Her calls became louder and more urgent until she sounded like an enraged chimpanzee. The male, smaller and with a more muffled voice, followed behind at a distance. As the calls continued, the female became increasingly incensed. But even after ten minutes, neither owl had come very close. Removing owls was not an easy business.

* * *

Weeks after returning home, I still wondered about the ethics of shooting something as charismatic as an owl. It seemed like an extreme form of wildlife management. The previous summer at a Forest Service research station in Oregon's Willamette National Forest, I had been on a guided walk with a veteran of eight years of fieldwork with northern spotted owls. Tim Fox had explained how he located spotted owl nesting sites. He spent hours stumbling across damp logs with a bucket full of white mice in his hand. The mice were the key to getting the owls to reveal their nests.

Fox had taught himself a pitch-perfect spotted owl call, which he sounded out into the forest as he walked through suspected nesting territory. If he heard an owl, he would hurry in the direction of the call. Spotted owls are hard to find, but when you do, they are not shy. Fox would often arrive to find an owl looking curiously at him from a nearby branch only ten or fifteen feet away.

The next part of the game was to set a mouse on the forest floor or on top of a log. The spotted owl, always looking for easy prey, would swoop down and grab the mouse in its talons before flying back in the direction of its nest. Fox now needed the nimbleness of a cat. He hurried through the undergrowth, keeping his eyes on the owl as it flew high into the canopy and disappeared into a hole or a crack in a giant Douglas fir or cedar. Fox would mark the spot on his GPS as he caught his breath. Sometimes, he would put another mouse at the foot of the nest tree just to be sure he had the right one. The spotted owl would step out of its nest and drop vertically, feet first, down the front of a 120-foot tree, plunging like a stone. As it neared the ground, it would puff out a few feathers to slow its descent, before landing right on top of the mouse. The owl wasn't flying. It was parachuting.

Fox had given several years of his life to the spotted owl. He clearly loved them. So when we asked him whether he agreed with killing barred owls to save the spotteds, it was a surprise to hear him say no. It was too high a level of intervention, he said. He thought it unlikely to work. And he wasn't too keen on the idea of killing a bird for doing what it was hardwired to do. Barred owls were simply exploiting a new niche. Despite all his work for spotted owls, Fox was ready to accept their fate at the hands of the barred owl.

A former employee at Montana's Owl Research Institute I spoke with was also dubious about lethal management. "I don't necessarily buy the argument that barred owls are not supposed to be there," he said. "I certainly get the sentiment to aid the recovery of the spotted owl, but I tend to err on letting things play out more naturally than that. It doesn't seem that removing one species to save another is an effective long-term solution." Barred owls and spotted owls, he pointed out, are also closely related. When the first few barred owls arrive in spotted owl territory, they tend to hybridize with the spotteds to create "sparred owls." The hybridization usually stops as more barred owls arrive in the territory. If the owls are that similar, how much does it matter if one replaces the other?

The two specialists may have had a point. In an ideal world, wild animals live wild. But here was a case where a heavy dose of management appeared necessary to keep a vulnerable species alive. The spotted owl was now, in today's lingo, "conservation-reliant." It was clear it needed help. But did this justify taking a shotgun to the barred owls? If so, wouldn't this sort of management threaten to take the *wild* out of the wildlife?

The ethicist in me took a step back to consider how the case stacked up. There are several conditions to meet if the argument for shooting one owl to save another is to hold water. First, you needed to be extremely confident the villainous owl is responsible for the decline. The experts at the Owl Research Center had warned me owl populations are notoriously difficult to track. Owls are mostly nocturnal and often live in hard-to-reach places. Specialist owls such as northern spotted owls or snowy owls can fluctuate wildly alongside booms and busts in their prey species. Getting a good count requires consistent and accurate fieldwork over many seasons.

When I put the counting question to Dave Wiens, he acknowledged there were challenges but said the barred owl case was clear. "We are extremely confident about their growth rate as a species. The range expansion is a huge, powerful event on a continental scale, expanding from eastern North America, across the Great Plains, into western North America. We are able to monitor their populations there very well." Ironically, many of the most convincing studies about barred owl numbers come from long-term spotted owl studies. The spotted owl is one of the most well-researched birds on the planet. Biologists looking for spotted owls always made a note when they encountered a barred owl.

"When first detected in the mid-sixties to mid-seventies in the Pacific Northwest," Wiens said, "barred owls remained at low populations. As these long-term spotted owl studies continued, they really saw an increase in the number of barred owls. They had an exponential growth rate in the Pacific Northwest." The bigger, feistier barred owls chased away the spotteds. The rise and fall of the two populations overlap perfectly. So the first condition about the barred owl's responsibility for the problem seemed satisfied.

The next condition for the ethics to work is to have confidence the barred owls arrived in the Pacific Northwest as a result of human influence. If they made their own way, it would be hard to justify intervening in the natural expansion of a species. It would be nature at work. Barred owls were originally confined to the East Coast because the Great Plains formed a barrier to their westward movement. There weren't enough trees to provide nesting sites for the several generations of owls it would take to expand across the country.

There are two candidate explanations for how barred owls overcame this barrier. The first is that natural swings in climate during the Pleistocene moved the Canadian forests far enough south for barred owls to do an end run around the Great Plains. A northern arc could have given them the trees they needed to move cross-country before dropping them back into Washington and Idaho after they passed the Rockies.

The problem with this explanation, says Wiens, is that the two species had a couple of million years to take advantage of fluctuations in forest cover throughout the Pleistocene. Despite the available windows, they didn't do it. "We know that these two species have been separate for a long, long time," Wiens said. "If it was climate change, barred owls would have had the opportunity to move earlier."

The alternative explanation for their migration is that human settlement of the Great Plains created the conditions for barred owls to hopscotch their way across the country. "As the settlers were moving across Great Plains," Wiens said, "they were planting tree belts." They also trapped beavers along rivers and creeks. With fewer beavers, streamside willows, cottonwoods, and other vegetation grew thick enough to support an owl migration. Fire suppression also allowed trees to grow taller. The barred owls could ride the settlers' coat tails across the Great Plains. "It is a pretty big coincidence that the migration happened at time of settlement," said Wiens. "You look at the range expansion and when it occurred, and it is pretty telling."

The science was not airtight, but it seemed to favor Wiens's account. The human influence was part of what made Hunt, Wiens's star remover, feel such remorse during her work. "All the evidence shows they are here because of human causes. It's more our fault than theirs," she told me. "They are just doing what they have always done. It's too bad for the spotted owl they are so good at what they do."

So we know barred owls are booming and causing a decline in spotted owls. It is also highly likely humans are responsible for their arrival. The remaining piece of the puzzle is whether killing them can actually solve the problem. This is what Wiens's study was designed to find out.

In the Wenatchee-Okanagan Forest, Hunt had no doubt. "It works," she told me without hesitation. She had seen barred owl numbers decline in her treatment areas. The evidence from her perspective was clear. Hunt's biggest concern was that five years of hard work would be undone if a long-term management policy was not implemented soon. She knew firsthand the barred owls came back immediately if you stopped shooting them. "It will be like the removals never happened," she said.

Wiens's report, filed six months after the experiment finished, supports Hunt's anecdotal account. The report concluded that shooting invasive barred owls "had a strong, positive effect on survival of native spotted owls." They found spotted owls stabilizing where there was barred owl removal and continuing to decline in control areas where they did nothing. In the areas Hunt staked out in Washington, barred owl populations at one point in the study had declined by 60 percent. Wiens told me the next step was to brainstorm a long-term policy for barred owl management with the U.S. Fish and Wildlife Service.

"Are there any options that don't involve killing?" I asked.

"There's a few," Wiens replied without enthusiasm. You could shoot birds with a drug to sterilize them. But that's even harder than shooting them with a twelve gauge. Some people talk about oiling barred owl eggs. "But to do that, you have to find the nest, which is exceedingly difficult," Wiens said. "Habitat management is also an option," he continued. "We have found subtle differences in how spotted owls and barred owls use the forest. One of the more interesting things is the vertical structure. Spotted owls tend to use the canopy layer and focus on canopy prey species like flying squirrels. Barred owls are more focused on the lower layers of the forest. Barred owls don't like areas with really dense understory. Spotted owls are more indifferent because they spend more of their time in the canopy." If you kept the understory thick, you could give spotted owls an advantage. But Wiens was also doubtful whether this strategy was practical. It would mean an awful lot of habitat management to marginally improve the odds. And the odds weren't good. "In reality," Wiens said, "maybe that would work if their population sizes were more equal, but now barred owls simply swamp out spotted owls."

Wiens had obviously thought through all the angles. Both he and Hunt had come to the same reluctant conclusion. It was ethically justified to kill one owl to save another. But it still didn't sound like fun leading such a controversial study. I asked Wiens how it felt to be the flagbearer for this work. "Certainly, there are a lot of people who don't hesitate to tell me what I'm doing is wrong. This includes scientists who say any kind of killing is not going to be the answer to anything," he said. After a pause, Wiens explained how he rationalized taking up the role of assassin in an ecosystem.

"Being an ecologist and studying predator populations for most of my career, in nature things work a little differently. Predation and apex predators have a large control over what's going on in these natural communities. What I see is humans using predation within a management context to maintain biodiversity. What we have is quite a powerful tool. The effects are immediate." Speed is important because northern spotted owls don't have much time. The situation is becoming desperate. The government in British Columbia has started to discuss capturing the province's last few northern spotted owls to breed them in captivity.

So humans, I put to him, study the situation and try to make a difference by behaving as if they were part of the ecological system. "It's not quite like predators," Wiens conceded. "We are not eating any barred owls. But the effects we are attempting to achieve are quite similar. Humans play that role with all kinds of other species. It's just not in our face like shooting barred owls is." When I raised this with Hunt, she pointed out the U.S. Fish and Wildlife Service conducted lethal management all the time. They have been killing cormorants for years to protect struggling salmon. Arctic foxes are culled for the benefit of a rare duck known as a Stellar's eider. Oregon, Washington, and Idaho have all received permits to kill sea lions preying on salmon congregating beneath the region's dams.

The obvious flaw in Wiens's analogy is that nature's predators don't have shotguns and tend not to kill things they have no plans to eat. Even if the analogy was ecologically grounded, it risked making the person who offered it sound a bit cold-hearted. But you don't have to talk to Wiens or Hunt for very long to realize this clearly isn't the case. "I really want to emphasize that I do truly grapple with the ethical side," Wiens said. "I think there are a lot of arguments that stepping back and putting your hands up and saying, 'Well, we can't do anything about this' has ethical consequences that in my mind could be a lot bigger." Doing nothing and watching spotted owls disappear could be at least as callous. Emma Marris, author of *Wild Souls: Freedom and Flourishing in the Non-Human World* and a well-known environmental science writer who wrestles with these dilemmas, has warned that doing nothing while watching a species slide to extinction leaves "blood on our hands." This was blood that Melissa Hunt and Dave Wiens, by killing barred owls, were both trying to avoid.

The conundrum in the Okanagan-Wenatchee Forest is becoming more and more common in recovery contexts. Human activities are implicated in the decline of so many species. Given this culpability, isn't there a strong obligation to help them make a return? And doesn't this obligation sometimes involve interventions that seem highly unnatural?

Perhaps there was a time when leaving animals alone was their best option. Perhaps that time may return. But in the interim, for some species, it might be necessary to wade into the system to help them survive.

* * *

I checked in with Melissa Hunt a few months later when her field season had wrapped up for the last time. She was getting ready to interview with Idaho Fish and Game for a more permanent job. She missed the owl project. She worried again about the time it was taking to reach a decision on how to manage the barred owls. All her hard work would be erased if nothing else was done. The barred owls were still reproducing and pouring into spotted owl territory.

I asked if she had any reflections on the ethical dilemma she had lived every night for five years now that she had a few months' distance from it. She looked up and spoke as earnestly as I had heard her speak. "This isn't something we are doing just because we want to," she said. "It is something we are doing because all the data is pointing to the fact these barred owls are a huge problem."

Hunt knew how the project looked from the outside. "It is important for people to take their emotions out of it and look at the data," she said, glancing down to the ground as she put her case together. I might have been overinterpreting, but I thought I detected a moment of hesitation. Science and her empathy for one of the forest's most successful inhabitants collided for a moment. But then science won out, and she looked back up.

"If we want spotted owls, it is something that is going to have to take place. It's not pretty . . . but one of those necessary things."

2 PONIES AND POLO MINTS

Stan Smith's Mazda Miata was parked at the forest gate when my sister and I showed up at Blean Woods in the county of Kent southeast of London. Smith is young and charming. The fleece, worn jeans, and work boots you expect on a conservationist were nowhere in sight. A photogenic smile, a thick head of light-brown hair, and chino-type trousers felt more Soho House than David Attenborough. Smith greeted us with enthusiasm, taking us straight to a map at the woods' entrance to explain what will take place here over the next two years. Then he quickly led us off down a forest track into the mottled light of one of the largest patches of ancient woodland in southern England.

Smith is the Wilder Landscapes manager at Kent Wildlife Trust. The centerpiece of his work is a project known as Wilder Blean, an effort to restore 1,250 acres of degraded woodland between the cathedral city of Canterbury and the English coast. The wood is a rare remnant of intact forest that has stood for more than four centuries. Today the oak and beech woodland is owned by an assortment of conservation organizations that all share the goal of turning a place more known for its links to Chaucer and Christian pilgrims into a hotspot of biodiversity. The woods have promise as a future home of large grazers, rare birds, and dwindling insects. Implementing these changes on southeast England's manicured landscapes requires management as intensive—if not quite as bloody—as the management of owls I had just witnessed in America's Pacific Northwest.

The eye-catching headline of Wilder Blean is the reintroduction of an animal more associated with the work of Paleolithic cave painters than this gentle corner of England. Smith is the public face of a project to bring European bison back to where they have not roamed in millennia. It's a complicated endeavor that will require considerable wildlife management skills to succeed. There is more than one noteworthy fact about the level of management involved. Let's start with this one: Smith not only plans to bring bison back to a forest where they have not roamed for thousands of years. He plans to bring them back to a forest where they may never have roamed at all.

* * *

Bison bonasus, like *Bison bison*, is the largest terrestrial animal on its continent. Males weigh up to eighteen hundred pounds and stand with the solidity of an ancient oak. Encountering a herd in a dark forest made generations of woodsmen pause to draw breath. Wisent, as they are also known, look different from Plains bison in North America, being a little less shaggy and lacking the giant shoulder hump. Their horns are more curved and point forward rather than up. They stand more erect than American bison and have proportionately longer legs.

European bison are a keystone species with many of the same ecological characteristics as their North American counterparts. They create wallows with their rolling, their hooves aerate the soil, and their fur provides songbirds with fluff for their nests. Their powerful tongues pull out clumps of vegetation and create a tussocky mosaic in the understory, adding a host of microhabitats for insects and small mammals. They browse as much as they graze, which means you will see them in forests with their heads up, nibbling bark off a tree trunk, or even standing on their hind legs reaching for leaves and acorns. Like their trans-Atlantic cousins, European bison graze in open country as well. They helped maintain grasslands in Europe during the Pleistocene, staving off the encroachment of forests with their powerful tongues and cement mixers of a mouth.

Bison bonasus in Europe had an even closer brush with extinction than the *Bison bison* in North America. By the First World War, the species was already struggling. Poland's Białowieża forest had the largest remaining heard, with seven hundred of the beleaguered beasts hiding in its depths. Exhausted soldiers and starving poachers made short work of the high-quality meat during the fighting. Some of the last wild bison were shot in Poland in 1919. A couple of stragglers were killed in the Caucasus Mountains in 1927. At their nadir, only fifty-four wisent remained alive in the world, all of them in zoos. Eleven bison removed from the Białowieża and a single bison from the Caucasus became the foundation of a recovery program. All European bison alive today are descended from these twelve animals.

Reintroductions started in 1951, and the population began to build. Painstaking effort goes into moving animals around to ensure maximum genetic diversity in the growing herds. A pedigree book published every year since 1932 keeps track of the lineage of every European bison. An individual bull might be trucked from the Netherlands to Ukraine, or two females moved from the Białowieża Forest to Bulgaria. The population has climbed to more than 7,500, of which four thousand live wild, with strongholds in Poland and Belarus. The Białowieża Forest has the largest free-roaming herd with a thousand animals. Wisent also roam semiwild in the Southern Carpathians in Romania, in the Rhodope Mountains in Bulgaria, in a state forest in Slovakia, and in western Ukraine. More tightly contained populations can be found in Germany, France, and the Netherlands. Very soon, under Stan Smith's watchful eye, they will be joined by a small herd in the unlikely setting of Kent's Blean Woods.

* * *

The story that circulated in the media when Wilder Blean's project went public declared *Bison bonasus* had been gone from the UK for six thousand years. In fact, Smith said, there was no evidence European bison had ever been there at all. Steppe bison (*Bison priscus*) were definitely once present. Thirty-thousand-year-old fossils proved it. "We just don't know that European bison ever got here." European bison evolved after steppe bison bred with aurochs during the late Pleistocene. The hybrid animal, identified through DNA evidence and the study of cave paintings at Lascaux and Pergouset in France, eventually became the only bison species on the continent. The best evidence suggests they never made it to England.

Undeterred by this minor detail, the Kent Wildlife Trust wants bison for an experiment in forest rejuvenation. If it works, it will serve as a model for other parts of the UK. Smith is keen to stress the experiment has as much to do with soil health and woodland ecology as it does with the shaggy, horned beasts that have brought such a flood of attention to Blean.

"For us it is all about ecosystem function," he said. "It would be fantastic if we could have steppe bison back, . . . but we can't. So we have looked for whatever the closest living animal is. And that is the European bison." The ecological need had been growing more dire as time went on. "We are not getting the levels of natural disturbance and mixed grazing pressure that we need to support these woods," Smith said. Without a large herbivore to chew on a few trees and keep the meadows open, wildlife in the woods had cratered. The previous owners of Blean added to the problem by converting 40 percent of the forest into densely packed, non-native Corsican pines. The grasses and shrubs that belong in the understory of an English broadleaf forest had withered and died, creating something close to an ecological desert.

Smith's employer had tried to nudge the system back in the right direction by selective cutting. But cutting was expensive, and it did not bring the full range of benefits of a more natural solution. They had tried introducing a few Konik ponies to see if their nibbling would help. But Blean's dog walkers started feeding the ponies carrots and polo mints. The ponies quickly became far too tame and failed to provide any of the benefits the woodland needed.

But *bison* in southern England after thirty thousand years?

"It's a bit mad," Smith conceded, stroking his chiseled beard as we stopped to look at some sweet chestnut coppice. "We are very careful how we talk about this. We try not to refer to it as a reintroduction. It is all about process. It is a nature-based solution to try to reverse the ecological crisis we are currently in. European bison are the best tool for the job." Several times, Smith mentioned the word *bioabundance*. He didn't want just a few bison and some healthier trees. He wanted to steer the woods back to a condition where they were awash with life again.

Picturing a bison stepping out from behind a smooth-barked beech or a gnarled oak, it was easy to share the excitement. Bison would be an eye-popping addition to the Kent fauna. In a culture more accustomed to mole, toad, and badger from *The Wind in the Willows*, a two-thousand-pound herbivore was attention-grabbing, to say the least. But Smith remained business-like about the rationale. "They are here to do a job," he insisted. If Blean was to recover anything of its former glory, it needed substantial manipulation. Kent Wildlife Trust was confident bison would do this better than a conservation officer with a chainsaw. And they would do it without a paycheck.

The arrival of the first *Bison bonasus* was still ten months away. But activity on site was hotting up. Fencing materials had arrived, and interpretive boards were being updated. The Trust had recently hired the UK's first-ever bison rangers. Over a thousand people had applied. Smith's press release for the job mentioned the rangers weren't expected to have experience with bison. After thirty thousand years, actual experience seemed unlikely. Amid the frenzy of applicants, they did get plenty who had worked with rhinos, elephants, and lions in Africa.

Returning the biggest, hairiest mammal to roam the English woods in several millennia would take meticulous planning. It was the diametric opposite of wolves showing up on their own in Belgium. Here in Kent, nature would not be finding its own way. People would tightly control the recovery. This echoed the question I had confronted with the spotted owl in Washington state. How much management of wildlife is appropriate to assist in their recovery? And something else quickly became apparent the more I chatted with Smith. Even though the project was driven by its ecological merits, the forest wasn't the only thing Kent Wildlife Trust was hoping to shape in this pleasant crook of England. Smith had his eye on something much bigger and more consequential than the fortunes of a small herd of shaggy, cud-chewing beasts.

* * *

The UK is one of the most nature-depleted countries in the world. A newly developed Biodiversity Intactness Index puts it in the bottom 10 percent of countries for the animals and plants it has left. It is last among the G7 industrialized nations. The bison at Blean were going to provide an extraordinary boost for the country's slim roster of megafauna. But Smith's downplaying of the charismatic bison and his enthusiasm for woodland bioabundance were typical of a difference I had detected between European and North American recovery efforts. It's a broad generalization, but Europeans seem more likely to talk about restoring ecological processes than to champion iconic species. When I asked Jake Fiennes, conservation manager of Norfolk's innovative Holkham Estate, which species he was most proud of helping to recover, he demurred. "I don't want to pick one out," he said. "I think that would be grossly unfair to say *this* was my keynote species. I think the effects of subtle changes have had benefits on multiple species." Likewise, Wouter Helmer, the Dutch director of Rewilding Europe, expressed a similar sentiment to me in a Nijmegen coffee shop. When I pressed him about which wildlife recoveries in Europe most excited him, Helmer's fleeting scowl made it clear I had asked the wrong question. "Talk about floods. Talk about erosion," he said, gently correcting me. "We are trying to bring back processes. Nature should surprise you."

In some cases, the preference for ecological health over flashy animals stems from necessity. Some of the larger missing species in Europe may be impossible to bring back. Wolves may have made it to France and Belgium, but it is hard to imagine brown bears emerging from hibernation in England's bluebell-covered dells. The effort to reintroduce Eurasian lynx to the UK had stumbled, in part due to lack of consultation with farmers. In some parts of Europe, sights are set realistically low. The transformation of the landscape is so complete that nursing the system back to health requires focusing on the basics. "We are starting from a much lower baseline," Smith said. The right vegetation. More native insects. Low-level predation of small rodents. The controversial stuff could wait till later.

The UK, in particular, has a cultural chasm to cross when it comes to reintroducing big animals. The bison returning to Blean would be managed under the UK's 1976 Dangerous Wild Animals Act. The act was originally written to govern exotic tigers and pet pythons and turned out to be the most suitable place to fit bison in the country's legislative superstructure. (The other options were livestock and zoo animals.) Not only did this seem a mildly absurd way to think about a (semi-)native species, but it created a logistical nightmare for Smith. The Dangerous Wild Animals Act requires land owners to actively protect the public. The bison didn't just have to be fenced in at Blean. People also had to be fenced out. Kent Wildlife Trust planned one small electric fence to keep the bison from wandering out of their allotted patch of woodland. Then, outside the first fence, it was building another fence, nearly six feet tall and topped with barbed wire, to prevent people from getting too close. Government authorities had also stipulated elevated footpaths and wildlife tunnels so bison could pass under public rights of way without the risk of running into people. The most expensive part of the restoration by far was not bringing a handful of bison to Blean from Germany, Scotland, and Ireland. It was the construction of twenty-eight miles of fence and four footbridges to keep wildlife and the British public apart.

The idea people and animals must be kept separate is not just a curiosity of British law. It comes from deep within the culture. I know this from my own experience. When I left southern England for the Rocky Mountains three decades ago, I remember the unease I felt almost every time I left the city. In the Poudre Canyon outside Fort Collins, I was certain there was a bear or a mountain lion behind every bush. When I camped, I was awake for much of the night, heart thumping as I strained my ears at the snap of every twig. A raccoon in the car headlights seemed miraculous. The deer nibbling on plants in town fascinated me. And the moose were astonishing. The whole picture created a puzzle. How could I safely occupy the same landscape as numerous large wild animals? To live safely in rural America, I assumed, you had to be Annie Oakley. And I had no gun.

The missing wildlife in UK no doubt has some benefits. Life sways to a gentler rhythm. Cream teas, croquet lawns, and manicured English gardens do have a certain charm. But this had real consequences for Wilder Blean. "The major pushback we are probably going to have here from site users is not really to do with bison," Smith said. "It is to do with the fact there will be animals on site at all. People aren't used to meeting animals. Full stop. It's so weird."

* * *

Smith's observation about missing cultural know-how had been on my mind since earlier in the day when I made a stop at the Wildwood Trust near Herne Bay. I went there to talk about a different reintroduction that had caught my eye. What I found was another case of animal recovery that would be intensively managed and closely monitored. But I also learned, as Smith had hinted, that the careful manipulation needed for a successful restoration was not just ecological. It was also deeply cultural.

Wildwood is a forty-acre site just off the A291 road on the edge of Blean Woods. You access it through a light industrial park containing an auto repair shop and a place to buy caravans. When my sister and I tried to find it, we wondered if we had taken a wrong turn. There wasn't an animal in sight. Eventually, we reached a sign and followed it to a parking lot ringed by a tall fence. As we got out of the car, suggestive avian squawks ripped through the broadleaf canopy.

The trust is both a conservation organization and a zoo. At its two sites in the UK, it keeps a wide range of native English fauna for public display. Behind the scenes, it also has a highly regarded team of wildlife biologists who breed rare species for restoration. The animals on site included wolves, bears, otters, beavers, hedgehogs, and polecats. They also had a couple of bison. We did a quick spin through the displays, admiring the storks and feeling sorry for the bears. The bison, I suspected, were pretty bored in their muddy woodland plot. But it turned out they were ecologically active. I later learned they had killed all the non-native trees in their enclosure with their rubbing and nibbling. My reason to come here, however, was not bison and their role in woodland ecology. It was wildcats.

I sat down with Laura Gardner, Wildwood's director of conservation; Vicki Breakell, the conservation projects officer; and Adam Roberts, Breakell's deputy. We gathered around a picnic table under an umbrella of deciduous trees. It was a gray morning, and the table had the cold, slick feel of a wet European spring. Walkie-talkies on Gardner's and Breakell's hips occasionally burst into life, alerting staff that an animal needed feeding or an overeager child had lost a parent on the maze of trails.

Gardner came to Wildwood after managing a rare bird collection at Leeds Castle. She was hired for her track record of building successful partnerships. Wildlife reintroductions need partners badly. The British relationship to animals, Gardner told me, was highly paradoxical. "We are a nation of animal lovers, and yet we are one of the most nature-depleted countries," she said. The absence of so many creatures for so long meant British people had lost all cultural memory of being around animals. "We are completely disengaged from wildlife," she said. Roberts agreed. "The animals you see in Britain on a day-to-day basis are all pets," he said. "To break that cycle is very difficult."

Wildwood had recently garnered international headlines when it launched its campaign to restore wildcats. Wildcats are the last surviving native feline in the British Isles. They look like a large tabby with a bushy tail. Several hundred of them cling to life in the Scottish Highlands where they are known—I assume a little tongue-in-cheek—as the "highland tiger."

Breakell has a degree in biodiversity and conservation management from the renowned Durrell Institute of Conservation and Ecology. She helped with Wildwood's reintroduction of water voles, which were under threat from invasive American mink. Wildcats are needed, Breakell told me, to bring balance back into the system. They would add a midrange carnivore to the depleted British mix. Their preferred prey is rabbits, but they also eat voles, mice, shrews, birds, fish, and insects. Their presence promised a mini trophic cascade, a reshuffling of the wildlife hierarchy, similar to what happened in Yellowstone after the return of wolves. Wildcats would give foxes, now the dominant carnivore, a run for their money.

The native cats, Breakell explained, have been struggling. They are killed by dogs and cars, shot by gamekeepers, and die from eating poison left out for other animals. Their biggest threat, however, is not killing but hybridization. The Royal Zoological Society of Scotland ran a study that compared the DNA of today's wildcats to the DNA of specimens stuffed in museums. It found the genetics of contemporary Scottish wildcats to be thoroughly diluted by the DNA of domestic cats. Highland farms and villages contained enough feral housecats to wreak havoc on the wildcat genome, one litter of hybrid kitties at a time.

The study spurred a report the following year from the International Union for the Conservation of Nature (IUCN), which concluded wildcats were already functionally extinct in the wild. Ironically, the most genetically pure wildcats in Britain were now the hundred or so animals held in captivity at various parks and zoos. The only way the species was going to survive was through a captive breeding and reintroduction program.

Wildwood joined two other conservation organizations for a feasibility study. It started fundraising to build ten new enclosures to breed the most genetically pure wildcats it could find in zoos around the country. All of the UK's captive wildcats have authenticity scores—one for DNA and another for coat pattern (pelage). The plan was to gather some of the best at Wildwood and carefully breed them. When sufficient numbers of authentic wildcats have built up, biologists will release between six and eight animals per year into suitable forests. Preliminary studies suggest Wales is the most likely release site because it provides the best opportunity to keep the captive-bred cats away from feral animals.

Gardner and Breakell made it clear the animals would not be left alone to run riot. "It is not a question of throwing a large number of animals out on the landscape and letting them fend for themselves," Gardner said. The reintroduction would be a tightly controlled affair. The wildcats would be health-screened and vaccinated before release.

There would be a ton of postrelease monitoring. The biologists would strictly follow IUCN guidelines for translocations. Everything that happened after release would be evidence-based and adaptive. Humans, at least initially, would carefully steer the return.

Biologists did not expect the prey base to be a problem, given the shortage of natural predators in the British countryside. The wildcats' instinct to hunt would still be sharp despite being raised in captivity, just as it is with domestic cats. The biologists would be ready, however, for any eventuality. "You might do some supplemental feeding initially while the population establishes themselves or has to learn those behaviors to hunt and forage," Gardner said. Nothing would be left to chance. Like Washington's owls, the vulnerable species would be subject to intensive management. But if the recovery was to succeed long term, it would take more than careful monitoring and regular interventions to make it work. It would take a cultural, as well as a biological, shift.

It was Roberts, Breakell's deputy, who put the missing piece of the jigsaw most bluntly: "We are asking people to change their relationship with cats." Before being hired at Wildwood, Roberts had worked in Africa with lions, rhinos, and baboons. He also spent time in Cambodia with an organization that rehabilitated working elephants and arranged alternative income for their owners. During interludes from traveling, he earned his master's degree in conservation biology from the Durrell Institute, the place where Breakell earned her degree. When he finished his program, he spent a year as a wild food forager, supplying over 250 products to London restaurants run by Gordon Ramsey, Hester Blumenthal, and other celebrity chefs. Put it all together, and Roberts had a pretty cosmopolitan perspective on wildlife and wildlands.

"I think we are too comfortable in this country. We have been so distanced from nature for so long," he said. I asked why this had happened. "I think it would be interesting to look at the island mentality," he continued. "The fact that we don't have to worry about another country's animals roaming into the UK matters. In the Netherlands, they are not in control of what countries on their borders are doing with their wildlife. Wildlife don't adhere to borders. In the UK, we don't have that issue."

"It is a bubble-wrap culture," Breakell added. "People don't have to think for themselves about risk." The British love wildlife, she said, but only from the comfort of their armchairs. She was right. I remembered the English person I met in the Northern Rockies who had come to America for a week of hiking. After the first morning out, they had second thoughts about their plans and retreated to their holiday rental for the rest of the trip. The thought of an unexpected encounter with a bear, a wolf, or a mountain lion had proved more unnerving than they anticipated.

None of the three conservationists I met at Wildwood thought the immediate prospects for a change in attitude were very good. But they all knew projects like Wilder Blean and the wildcat restoration would give people the opportunity to witness more hooves and fangs on the ground.

"If you can have this demonstration project where people can come and see it, you are building people's trust and confidence," Gardner suggested. "The school groups we see, the conversations: it's really exciting. There such a momentum about rewilding and native species conservation." Over time, they all hoped, the baseline might shift back as British people reacquainted themselves with being around wildlife. For now, Gardner said, it remained a cultural minefield. Trying to persuade the English they needed more predators on the landscape threatened to put a real dent in their Englishness. It was like trying to persuade them to drink a little less tea.

* * *

A few hours later in Blean Woods, Smith and I stood over a seething mound of southern wood ants. "You can hear them if you listen closely," he said. I moved closer and lowered my ear toward the pile. The ants are described in the guidebooks as aggressive and have the ability to spray formic acid at threats. "If you stand still too long, they will crawl up your leg," he told me. I decided I could listen for the ants another time.

Insects are one of Blean's strongpoints. The oak trees are home to more than 350 species of invertebrate. There are 127 different spiders in the woodland, including one, *Walchenaeria mitrata*, which has been documented in the UK only nine times. Earlier, Smith had stopped to show us a verge of cow wheat, a semiparasitic plant used by caterpillars of the rare heath fritillary butterfly. Blean is home to a sizable portion of the UK's heath fritillaries. It also hosts a chunk of the UK's rare lesser spotted woodpeckers. It was the butterflies, spiders, and woodpeckers, not the bison, that had earned Blean national recognition as a Site of Special Scientific Interest.

Despite its obvious appreciation of the diminutive, Blean was doing its part to nudge the English toward the idea that larger animals are also essential. While the bison grabbed the headlines, other grazers would play key roles in Blean's recovery. The People's Postcode Lottery was funding a perimeter fence around the whole property. Exmoor ponies, a skittish UK breed, would replace the carrot-addicted Koniks. The Exmoors are expected to follow along behind the bison, grazing the grasses shorter. Longhorn cattle would approximate the ecological role of extinct aurochs, and iron age pigs would simulate the rooting of wild boar. Unlike the bison, who would be restricted to distinct sections of the woodland, the other grazers would roam free across the property, giving the scientists an opportunity for data gathering and the public the chance to interact with animals not categorized as dangerous or wild.

The longer-term vision included the reintroduction of pine martens, a medium-sized carnivore that feeds on small rodents, birds, and insects. A sprinkling of pine marten lives in Wales, northern England, and Scotland. A few also cling on in Hampshire's New Forest. Wilder Blean wanted to give them a second southern home in Kent. Native red squirrels might also be on the agenda and, possibly, a reinforcement for vanishing turtle doves. Beavers, another potential returnee, were already less than twenty miles away at Ham Fen, the site of the UK's first beaver reintroduction nearly two decades ago. Smith thinks there is a chance Blean might provide beavers a good home if the Trust digs a few more ponds and builds some strategically placed levees, known locally as *bunds*. He told me about ongoing work to reintroduce the red-billed chough. The chough is a striking species of crow with a blood-red bill and bright red feet. It is prominent on the city of Canterbury's coat of arms but has been extinct in the region for two centuries.

Excited as I was about what lay ahead, I had to confess I was wrestling with the large amount of hands-on management the reintroductions seemed to involve. Some of this was mandated by the IUCN. You can't just drop an animal into an ecosystem and wish it good luck. But some of it felt like a British overreaction to the whole idea of wild animals. There would be electric fences, vaccinations, and the possibility of supplemental feeding (for the wildcats). The bison would have clearings cut in the thickest part of the Corsican pine forest to help get them started. As time went by, some bison would be swapped out with fresh ones from Europe to maintain the herd's genetic diversity. Smith even mentioned the possibility of bison wearing satellite-guided shock collars recently developed in Norway to replace the fence at some point. If they stepped over the property line, they would be zapped.

I could understand the importance of these strategies for the success of the reintroductions. But wasn't there something to the idea of leaving wild animals alone? In his book *Wild Ones: A Sometimes Dismaying, Weirdly Reassuring Story about Looking at People Looking at Animals in America*, Jon Mooallem throws up his hands when he witnesses overintensive wildlife management. "Zoom out," he says, "and what you see is one species—us—struggling to keep all others in their appropriate places, or at least in the places we've decided they ought to stay." Mooallem has a point. A bison steered by a satellite seems to be missing something essential to its bison-ness. It has lost the intrinsic value of simply being wild. The same question had come up around spotted owls. Humans were heavily implicated in their survival, maybe permanently. But how far can you go with wildlife management before you lose something essential of what you are trying to save?

"Wild animals do gain something from being genuinely wild," Smith said, when I put this question to him. But he hesitated to frame the justification for bringing them back in terms of a romantic vision of species completely untainted by the human touch. "We have to work with the fact it's modified. We are never going to be able to recreate this idyllic, fictional past." The point about looking forward rather than back seemed right. Relationships with large animals on highly developed landscapes like the UK's have to be viewed in a future-oriented way. Reintroductions have to be hands-on. Conditions on the ground simply don't allow for it any other way. Part of me was reluctant to accept it, but I could see it would require the right sort of management—however intensive, however distasteful to the purists—if people and animals were to coexist.

Coexistence was certainly not new here. Blean Woods, natural as it seemed at first glance, was riddled with evidence of human use. In the middle of our walk, we stopped to look at the Radfall, a fifty-foot-wide swale built through the woods in the Middle Ages. The Radfall connected Canterbury to the fertile grazing grounds of the Kent coast near Whitstable at a time when wolves still chased deer through English forests. Its hand-built banks, several feet high, kept livestock from wandering into the woods and offered some protection as herders goaded their flocks coastward. Villagers also used it bring pigs to forage for fallen nuts, a practice known as *pannage*.

The management also extended to the flora. Sweet chestnut coppice, much of it now overgrown, let foresters manage chestnut stumps for fence posts, thatching materials, and hop poles. The chestnut had been introduced to England in Roman times. Smith pointed out twenty-foot-high sprouts ringing ancient stumps waiting to be cut. Over the centuries, the coppice had become good habitat for insects. The Wildlife Trust coppiced what it could afford for the benefit of the heath fritillary butterfly, an insect known for good reason as the "woodman's follower." The butterflies were dependent on the coppicing.

Villagers had used Blean in numerous other ways. Clay soils and ample firewood meant the area had been a center for the tile industry. The bark of the ancient oaks was ground up for a solution used by the tanners in Canterbury. Oak sawdust went to the coast for curing fish. Beech was used to make potash, a raw material in gunpowder. Hazel was woven into fences. Hop gardens occupied the clearings. Coopers, wheelwrights, shipbuilders, and coffin makers all earned a living from Blean.

Kent Wildlife Trust thought the extensive human history was, in some respects, as important to the woods as the wildlife. As well as hiring two bison rangers, it had recently hired two professional storytellers. The Trust wanted to show how human and natural history evolved here together. The forest and the people had shaped each other. And those two histories would remain entwined moving forward. "Our landscapes need to work for people," Smith said. "We can't just wall areas off and say, 'This is where the wildlife is.'"

In the blend of wildlife and culture Smith envisioned, however, animals had far greater need for an elevated profile. In England, Smith said, wildlife had been pushed too far into the shadows. "The kind of national parks we have in the UK are very much cultural landscapes. They are not wildlife driven," he said. He hoped Wilder Blean could tip the balance slightly back in wildlife's favor. Smith wasn't trying to recreate Yellowstone. But he did hope to ratchet the profile of big animals a tiny bit higher in a reluctant English culture.

When we got back to the vehicles at Blean's gate, I asked Smith if he thought he represented a new generation of UK wildlife advocate. Smith has that self-effacing manner of the English, but he seemed to recognize the significance of what's happening. "I think it is definitely new. Nobody has tried to do what we are doing. The younger audience seem more receptive to it. It's a big change."

I asked how comfortable he felt being one of the public faces of the change.

"I had not thought about it before," he said with an embarrassed laugh. "But we need to do it. It is necessary. It would be nice to just hide away from it and do some small project in the woods. But we need to be doing it in full view of the public and take what comes with that."

There would be no end of intrusive, hands-on management of wild animals taking place here in the next few years. It had to be that way if these woods and the species they needed were going to recover. But it was clear the project at Blean Woods was also an experiment in management of the human animal. I asked Smith if he thought an implicit part of his work was to force a change in British people's relationship with animals.

"Yes, absolutely," he said. "It is not necessarily a really comfortable position because you have to respect other people's opinions. But you are trying to change behaviors and attitudes." Wilder Blean wanted to move the needle on wildlife acceptance in the UK, a needle that has been jammed for a century.

The crew at Wildwood had made the same point to me earlier in the day. "Whether these projects are successful or fail, they are driving the conversation," Roberts had said. "A huge part is harnessing the power of media to really start talking about the future of nature. I think, animals aside, the most powerful thing about these projects is their ability to make people think a little bit differently about the environment around them."

So, in the end, there is plenty more at stake than bison, butterflies, and wildcats in these projects. The bison would nibble the trees and create habitat for heath fritillaries, rare spiders, and turtle doves. The wildcats would tuck into the UK's supply of rabbits. A number of dormant ecological processes would resume. The scientists would fret around the edges, trying to maintain some control, as they watched the ecological processes rev back up again. But here in the UK, the work of animal recoveries is set to transform the people as much as the ecology. The animals, of course, have spent millennia becoming experts at the ecological work. Those who dare return large beasts to the garden of England will have to solve the cultural challenges a little more quickly.

3 PRUNING APPLES FOR BEARS

I was struggling to keep pace with Mario Cipollone as he hustled up a steep trail in the central Apennines of Italy. We were heading through a cold rain toward an abandoned shepherd's cabin in the Monte Genzana Reserve. Fallen beech leaves made a tawny carpet as we crossed a wooded ridge at four thousand feet. Wolf scat by the side of the trail provided a clue these rugged hills concealed wildlife far larger than I was accustomed to seeing in England.

As we descended the other side of the ridge, the thirty-eight-year-old conservationist, his hair cropped short atop his muscled physique, scanned the dripping forest for something he wanted me to see.

"There!" he said, pointing toward a tree with a familiar look. "Apple! See how our volunteers have pruned it?"

The gnarled trunk was a relic from the time these hills were cultivated by Italian farmers, before they abandoned their fields at the end of the Second World War. Volunteers under Cipollone's guidance had removed the dead wood and opened the tree to the light to stimulate the growth of fruit. The apples from these harvests were not for the markets in nearby villages but for the bellies of one of the region's most notorious residents: the Marsican brown bear.

"We don't talk about these apples a lot," Cipollone confessed with a sheepish grin. "Rewilders aren't supposed to be pruning trees."

Cipollone and his partner, Angela Tavone, who joined us on the hike, work for Salviamo l'Orso, an organization devoted to protecting the highly endangered Apennine bear. As people have left this part of L'Aquila, big animals, including the elusive bear, have nosed their way back. Most Marsican bears reside within a network of national and regional parks. In recent years, they have increased in number and started to explore the abandoned farmlands and orchards on the edges of the picturesque villages. Cipollone and Tavone are helping smooth their return.

The work I witnessed on behalf of the Marsican bear is innovative and effective. A rare and charismatic animal is coming back from the brink. But pruning apple trees for the benefit of bears is an almost heretical type of intervention. Italy packs a lot of people into very limited space. Wandering around the edges of human habitation is a startling array of large animals. In the Washington forest and in Blean Woods, I had already seen wildlife managers being remarkably hands-on as they helped wildlife recover. Here was a different wrinkle on the same dilemma. How much intervention is acceptable when people and animals no longer have the space to live separate lives? Or, to put this in terms of what I had just seen, should Cipollone and Tavone be pruning apple trees to help grow bears?

* * *

The Marsican bear (*Ursus arctos marsicanus*) is one of the rarest of the world's fourteen or so subspecies of brown bear. (The exact number is still a matter of some debate.) The bears live at the geographical center of Italy, a country well-trampled by sixty million people. Their population has rarely climbed much above fifty since biologists started paying attention nearly a century ago. Thanks to recent efforts at coexistence, the bear's prospects are beginning to look up. Cipollone suspects at least eighty bears now live in and around their stronghold of Abruzzo, Lazio, and Molise National Park. From 2016 to 2019, ten or more cubs were born each year. In 2019, a record sixteen cubs were spotted. In 2020, a female (known to bear-watchers as "Amarena") showed up near the village of Villalargo with four healthy cubs in tow. Two yearlings made their way north to Gran Sasso National Park, a place where bears had not lived in a century. Slowly, tentatively, the bears are starting to rebound. And although the recovery remains precarious, Cipollone's sights are set high. "In this extended core area," he said, eyes sparkling, "I think 250 could live." His face broke into a huge grin. "I would dream to see this."

Cipollone grew up in the countryside near Pescara, where he spent hours walking the woods in search of animals. At the age of eight, a farmer taught him how to build a snare, hoping to nourish a kindred spirit in the boy. When Cipollone came home one day to find a neighbor's cat looking back at him from his trap, he was appalled. He swore off harming animals for good.

As a teen, cities bored him. The world beyond bricks and pavement held far greater allure. After he headed off to college and his knowledge of wildlife increased, Cipollone became convinced the Apennines harbored something special. Though depleted by centuries of hunting, big animals such as wolves, wild boar, and the goat-like chamois clung on in the region's remote corners. With the farming economy fading and the forests starting to grow back, Cipollone began to see wildlife as the key to the region's future. In 2012, he cofounded Salviamo l'Orso to address what he considered the cultural emergency created by threats to the iconic bear.

Italian zoologist Giuseppe Altobello classified the Marsican bear as a separate subspecies in 1921. During hundreds of years of isolation from the nearest bears to the north, they developed a distinctive lower jaw for breaking open the beech nuts and acorns that form a large part of their diet. In addition to the unusual jaw, they have a much shorter snout than brown bears in the rest of Europe. Bears outside of the central Apennines, which roam across twelve hundred miles of mountains and foothills from the Caucasus to the Pyrenees, look more like each other than they do their acorn-eating cousins halfway down the Italian peninsula.

When European sentiments about conservation began to stir at the beginning of the twentieth century, Italy was among the first countries to provide protection for its wildlife. The government created Italy's first national park, Gran Paradiso, in 1922. The same year, at Altobello's urging, the regional authorities set aside land in Abruzzo for the benefit of wolves, chamois, and the struggling bear. Altobello had just identified the Apennine wolf as its own subspecies and knew the bears and wolves were both teetering on the edge. The land eventually grew into the National Park of Abruzzo, Lazio, and Molise. Protection came just in time. The park's mountainous slopes were a vital refuge for dwindling species. Italians quickly grew to appreciate the park. The idea you could see a brown bear just a couple of hours after eating breakfast in Rome's Piazza Navona quickly became a source of national pride.

The Marsican bear is Italy's largest endemic carnivore. The big males weigh over four hundred pounds and tower six feet when standing on their hind legs. Despite being opportunistic carnivores, 80 percent of their diet is fruit, nuts, and herbaceous plants. In spring, the bears scour low-elevation woodlands and open pastures for plants and insects. In April and May, the hormone-charged males compete with each other for mates. As the summer sun warms the high country, the bears move up in elevation to chase a wave of new plant growth washing over the ridges. There they occasionally bump into shepherds who have brought their sheep to the mountains for the summer as part of the centuries-old practice of transhumance.

The bears scavenge meat when they can, occasionally taking advantage of the leftovers from wolf kills (a practice Cipollone called *kleptoparasitism*). Come fall, the hungry foragers return to the valley bottoms to comb the low country for fruit, beechnuts, and acorns. The fats and sugars in the high-calorie foods help them prepare for the months of hibernation ahead. The first snows signal the final weeks of foraging season, and the bears' thoughts turn toward their dens. As the seasons start to change, the fertilized eggs carried by the females finally implant into wombs made hospitable by the frenzied feeding. Some females start the process of gestation snuggling in the company of ten-month-old cubs from the previous season who join them in their winter hideouts.

Like many brown bears, the Marsican bear gives birth in the den in January or February. Mothers average between one and three young, occasionally four. The cubs are ten inches in length and weigh a pound or less at birth. The newborns lack hair and teeth and crawl blindly up their mothers' fur until they find a teat, the warmth of her body keeping them alive as snow falls outside. The burrito-sized young grow rapidly on the mother's rich milk. They emerge from the den in April still suckling and gradually gain a taste for the emerging plants and insects. As their confidence grows, they bumble around behind their mother, learning how to forage and play. They grow fast, though it is more than a year before the youngsters wean from their mother's milk.

As brown bears go, Marsican bears are surprisingly mild-mannered. Cipollone, Tavone, and the several dozen volunteers they recruit every summer don't take bear spray with them into the woods. When I expressed my surprise at this, Cipollone shrugged. "The genes for aggression have been weeded out over the centuries," he told me. With such a tiny breeding population, it makes a big difference which individuals survive. Hikers certainly have to stay alert, but generations of hostility toward Apennine bears have slowly crafted a gentler ursine.

The fact the bears are little threat to people does not mean the favor gets reciprocated. Since 1970, 124 Marsican bears have been found dead—a frightening number for such a tiny population—and 80 percent of them died at the hands of humans. Bears have been poisoned, accidentally drowned in large water tanks used by shepherds, and shattered by collisions with cars. Sometimes bears are shot by mistake during hunting season when the woods are filled with rifles, dogs, and trigger-happy men in pursuit of *cinghiale* (wild boar). It's a risky time for bears trying to fatten up for hibernation.

Some bears die from spite. In 2007, three bears (along with several wolves and wild boar) were poisoned within the boundaries of Abruzzo National Park. The chief suspect was an unhappy livestock owner. The bear has faced a long battle to win people's hearts. "We have to fight people's fear of wildlife," Tavone told me. "We try to bring people to our side in terms of the positive effects of the presence of wildlife." Not only were

Tavone and Cipollone wielding pruning saws on behalf of the Marsican bear. They were engaged in a full-on charm offensive.

* * *

Angela Tavone was born in Bojano, a town of eight thousand just beyond the southern portion of the bear's range. Her interest in the natural world evolved more gradually than Cipollone's. Her parents took her for the occasional picnic in the mountains but showed only a passing interest in nature. The first inkling of her love of wildlife came on a school trip to a historical site on the edge of town. When the history lesson was done, the teacher had the children take casts of animal tracks they found in the mud. Tavone recalls how astonished she was by a fox print. The idea there were wild animals padding around whose lives secretly crossed paths with hers was a revelation. She named a tree in her parent's garden "Betulla" and started retreating to its embrace on summer afternoons to read a growing collection of wildlife encyclopedias.

By her teens, Tavone was hooked. "When I chose the university," she said, "it was clear to me that discovering the natural world was my first interest." She went to the nearby University of Molise to study the science of cultural and environmental heritage. She volunteered in parks during the summer breaks, despairing whenever news filtered through of another bear poisoned or shot by local farmers. It was obvious to her the bear had a public relations problem. Eager to help, Tavone chose graduate degrees in science communication and conservation governance. For her dissertation, she traveled six thousand kilometers through most of the protected areas in the Apennines to investigate how the parks presented their mission to the public. When Salviamo l'Orso needed an outreach specialist to help spread the word about the recovering bear, Tavone was a perfect fit for the job.

I asked how she spent most of her workday, and the answer was quick. "Writing!" she said, with a hint of exasperation. "I'm writing posts for social media. I do interviews and try to combine good images with positive messages." She also submits grants to the European Union and supportive foundations to fund Salviamo l'Orso's operations. She uses her savvy in science communication to build a case for the region's future as a wildlife mecca. Step outside the stone walls of the medieval towns, Tavone says, and Alaska-sized wildlife amble through the woods. The secret weapon she deploys to make the case has more than enough charisma for the job. "Most people love the imagery of these bears," she says with a smile. Marsican bears stroll regularly through Facebook feeds around the world thanks to her nimble fingers on the keyboard.

Salviamo l'Orso's work took off in 2015 when Rewilding Europe created Rewilding Apennines and hired Cipollone and Tavone as team leaders. Rewilding Europe's goal is to restore ecological processes to the landscape. The organization has spent more than a decade scouting areas where falling human populations and a need for new economic opportunities have created potential for rewilding. The Central Apennines is one of eight regions—others include the Danube Delta, Portugal's Côa Valley, and the Southern Carpathian Mountains—that serve as showcases for European rewilding.

From their Dutch headquarters, Rewilding Europe directs funds toward promising recovery projects in the target regions. Its long-term vision may be an idealistic one. But it brings a slick publicity machine, a proven ability to raise money, and a pragmatic sense of how to get things done. The organization makes a business case for the wild, and its enterprise team works with local entrepreneurs to identify economic opportunities like artisanal cheese production and wildlife lodges. A dedicated loan fund, Rewilding Europe Capital, helps nature-oriented entrepreneurs get started.

In the Central Apennines, Rewilding Europe found a robust system of parks connected by thousands of hectares of abandoned farmland. Holding things together was a fragile network of environmental organizations run by experienced activists. In charming villages surrounded by forested hills, Rewilding Europe found communities hungry to revitalize their economies. What it also found in the Central Apennines was wildlife. Plenty of wildlife.

The creation of Abruzzo, Lazio, and Molise National Park in the 1920s provided a lifeline for more than just the bears. Apennine wolves and the native wild boar also took advantage of the new park and stabilized their numbers. When the government added a series of national and regional parks in the 1980s and strengthened laws against poaching, the recovering animals started to spill out beyond park boundaries. Often, the animals did it by themselves. But sometimes, like the owls and the wildcats I had encountered elsewhere, they needed a hand. In addition to the Marsican bear, a mountain goat, a deer, and a giant, blond-headed vulture are some of the notable beneficiaries of biologists' efforts. And I didn't have to go too far from Monte Genzana's apple trees to see some of these helping hands at work.

* * *

The princely chamois live best amid crumbling sedimentary rocks high in the Apennine range. Their curved horns and painted faces crown bodies taut with muscle. Chamois graze on lichens and grasses and retreat to inaccessible cliffs at the first hint of trouble. Their cloven hooves are specially tailored for the terrain, with rubbery central pads to provide a sure grip and hard edges to give purchase on small protuberances of rock. A spooked chamois can defy the laws of physics when it bounds at thirty miles an hour over territory on which most humans could barely cling with a rope. Overhunting in the nineteenth and early twentieth centuries took its toll, and by the 1920s, Abruzzo was the only place in the Apennines where chamois still clung on. After the creation of the park, the nimble goat's population stabilized and slowly began to grow.

The species were out of immediate danger but were far from recovered. Since chamois stick to the high country, they were unable to cross the valley bottoms to expand into neighboring ranges. A series of reintroductions begun in the early 1990s relocated dozens of chamois from their Abruzzo stronghold to other parks. One of the targets for reintroduction was Monte Sibillini National Park, 125 kilometers to the north of Monte Genzana. Not long after meeting Cipollone and Tavone, I traveled to Monte Sibillini to find out how the reintroduction was going.

The diesel minibus cut south from highway 77 in the central Italian region of Marche. We were only a few minutes off the highway when the bus starting sweating its way upward towards the remote Sibillini park. The road grew progressively narrower as it threaded its way through a rumpled blanket of hills toward the village of Bolognola. A smattering of hamlets and farms languished on the roadside in the early June heat. When we pulled up in front of Bolognola's church, the stifling air was freshened by the sound of water tinkling from a village fountain. Voices from the cafe across the street and the occasional swing of the cafe door completed the sleepy scene.

Sibillini National Park biologist Alessandro Rosetti met us in the village square. A slender man in his late forties, with unruly curls of black hair, Rosetti was happy to see visitors. The region was still recovering from the effects of a series of earthquakes in 2016 and 2017. Several of the stone homes in Bolognola were wrapped with braces and steel support cables to prevent them from collapsing. Rosetti and his family had been unable to return home to nearby Visso since the quakes because of the risk of further collapse. The tourist economy across the region had cratered.

After some friendly chit-chat and a quick dive into the cafe for some much-needed liquids, Rosetti walked our small group up a chalk path toward the chamois enclosure. On the way, we passed a monument to thirty-four villagers who had perished in two avalanches in the 1930s. Back then, the surrounding mountainsides were denuded by logging. Slabs of unstable snow often raced down the steep slopes. Half a century later, the villages contained fewer people, and the forests had begun to grow back. The recovering trees created a blanket of security for the villages below.

Arriving at the enclosure site fifteen minutes later, Rosetti told us to stay back while he walked to a covered area where the biologists provided food for chamois acclimatizing to their new home. The agile mountain dwellers spend several weeks adapting here before they are transported across the valley to their permanent home on the surrounding cliffs.

Rosetti banged on the side of the trough and shook out some hay before retreating outside the enclosure. We waited ten minutes before a pair of brown goat-like animals with short black horns carefully picked their way down the steep hillside from above. These were two of only a handful of chamois still kept in the enclosed area. More than 250 were already firmly established on the Monte Bove massif a few kilometers away. The wild herd was now big enough to take care of itself. Rosetti told us the enclosure was essentially an educational sideshow at this point. The chamois still produced young. These were useful as backups in case anything changed for the animals on the massif. But the Sibillini herd as a whole no longer needed the help. Today, upward of three thousand Apennine chamois dust the mountaintops of several parks in the region. The biologists who conducted the reintroductions have retreated into the background again, satisfied with a job well done.

Chamois are not the only species in Monti Sibillini that have recovered in the last few decades. Seventy-nine red deer brought in between 2005 and 2012 have grown into a herd of nine hundred. Wild boar have achieved a similar feat on their own. There are now four times as many of the bristly porcines as ecologists think the beech and oak forests of the region can bear. It's a similar story of profusion with wolves. The Apennines' dominant canid enjoys some of the highest pack densities in Europe. Studies show boar make up two-thirds of the diet of the returning wolves, something all parties—except the boar—are happy to see.

The conservation successes in the region are not only on the ground. Riding the thermals that mix the air above the limestone peaks are growing numbers of Griffon vultures. A sure sign of improving wildlife is the return of a species that eats their carcasses. These twenty-pound scavengers had been eradicated from Italy, with only a few surviving on the nearby island of Sardinia. Farmers mistakenly thought them a threat to their livestock and spent decades in a misguided campaign to poison them using carcasses laced with strychnine. The vultures were in fact no threat. They are scavengers, not hunters. Their association with the recently deceased is so tight that a group of vultures at a carcass is known as a *wake*. But the misunderstanding cost the birds dearly. They were poisoned off the Italian peninsula, finding refuge only across the sea in North Africa and Spain.

With protections in place, Griffon vultures are flexing their nine-foot wingspans all around the Mediterranean again. As populations rebuilt, vultures spread first to Sicily, then the Apennines, and then the Alps. Five hundred of the ancient birds now soar over the Central Massif in France. The Spanish population has exploded from a low of 3,500 in 1980 to about 25,000 today. A crackdown on poisoning and a better design for electricity pylons have yielded steady rewards.

Griffon vulture recovery, like Marsican bear recovery, energizes wildlife enthusiasts but raises questions for those who prefer a hands-off approach to wildlife. Their return has involved a substantial entanglement of vulture and human lives, reflecting the often complicated relationship between wild animals, people, and livestock.

When wildlife declined in the last century from overhunting and trapping, vultures came to depend more and more on dead livestock for food. Sheep, goats, and cattle that died in the mountains became an important source of calories. This stopgap for the vultures turned out to be highly beneficial for farmers. The strong acids in the vulture's stomach were ideal for neutralizing the diseases carried by livestock. For a time, the vultures became valuable—if underappreciated—sanitation workers.

When vultures started succumbing to the farmers' mistaken poisoning campaigns, livestock diseases proliferated. Farmers were instructed to burn dead livestock to keep diseases at bay. This may have helped the farmers, but it meant that when vultures started to recover, they found hardly any carcasses to eat. Biologists had to step in to maintain the recovery. Artificial vulture restaurants—stocked with dead horses and roadkill hauled in by volunteers and state agencies—helped provide a bridge in the fragile recovery. Vultures recovered successfully, but only because concerned people went to great lengths to feed them.

Going the extra mile for avian scavengers turns out to be a common practice. A different kind of intervention has been required for California condors in Arizona and southern Utah. The condors—reduced to a low of twenty-three in 1982—have been making a slow but steady recovery through the release of captive-bred birds. Like the Griffon vulture, the California condor lives by scavenging. Many of the carcasses they consume are riddled with lead shot from hunters who injure animals or fail to pick up their quarry. The condors die slow deaths from lead poisoning.

California has banned the use of lead bullets, but other states are less proactive. Biologists are trying to mitigate the problem by capturing the wild vultures and treating them with chelating agents. A poisoned bird needs this therapy twice a day for five days to purge the lead from their blood. The treatment removes the toxin and has helped keep the recovery on track. It is a labor-intensive process that's possible only because so few condors live in the wild and so much money has been invested in their recovery.

These efforts on behalf of California condors reflect how the species is highly dependent on humans. At this point, they are conservation-reliant and need the intervention to survive. The strategy has echoes of Cipollone and Tavone's pruning of apple trees for bears and Hunt's shooting of barred owls for the sake of the spotteds. In an ideal world, none of these interventions would be necessary. Wildlife would be left to their own devices. But some recoveries are so tentative that extraordinary measures are necessary. It is not clear when—or even whether—the helping hand will be withdrawn. The situation is especially complicated when animals are returning onto thoroughly humanized landscapes like Italy's crowded mountains or the Kent woods. Wild animals don't get their own wildernesses anymore. They return to places they must share with humans.

A more permanent solution for vultures lies in the continued recovery of wild animals, together with strong laws against the use of lead shot. Vultures need clean animal carcasses and plenty of them. As chamois, red deer, and wolf recovery continues in the Apennines, the vulture's prospects will improve. More wild animals living alongside more sharp-fanged predators is good news for scavengers. It means better prospects for the region's aerial clean-up specialists.

* * *

The rain had stopped by the time I walked toward the apartment in Pettorano sul Gizio where Tavone and Cipollone live nine months of the year during bear season. Their house was in a row of attractive stone and plaster buildings on a street near the top of the medieval village.

From a compact kitchen that doubles as an office, the young couple looks east over a creek toward a forested slope. If they are lucky, they might glimpse a Marsican bear from the kitchen window as it ambles across a distant clearing.

Pettorano sul Gizio is becoming an increasingly popular place to live. Word-of-mouth has turned the village into a destination for Europeans interested in rewilding. Several dozen university students come from across Europe each summer to volunteer on behalf of the bear. Artists have visited for residencies. A team from *National Geographic* filmed a documentary, and a renowned Italian photographer has relocated here. The historic castle in Pettorano recently had an upgrade in anticipation of more visitors. The castle's curator greeted Cipollone and our small party warmly when we stopped in for a quick look. There is a sense that rewilding is a new opportunity that has come the village's way. The winner of the latest mayoral election is the former director of Abruzzo National Park and the current chair of Rewilding Apennines. All of this is good news for a rustic village whose population has just started to grow again after decades of decline.

We had dinner that night at a restaurant near the top of the small knob on which most of the village sits. "The owner picks mushrooms for his menu from the woods around here," Cipollone told me, as he gestured me to enter through the restaurant's large barn doors. We entered a cozy room with stone arches and an imposing wooden bar. Everything smelled of candles. Tavone and Cipollone went out of their way to make small talk with the *proprietario* about his seasonal mushroom harvest. They knew they would run into him in the woods. They also knew he was irritated by restrictions on firewood gathering issued to protect the bears. But it seemed to me they were winning him over. He appreciated they were bringing people to his restaurant outside of peak summer season.

While I tucked into pasta with asparagus and mushrooms chased by a glass of local wine, we talked more about the ups and downs of bear conservation. Cipollone is not convinced every "accidental" shooting of a bear by hunters over the years has been an accident. Although a majority of the population of Pettorano sul Gizio seems supportive of their work, there are still people that need to be convinced. Not everyone is excited about an Italian version of a grizzly bear showing up in their backyard.

Tavone's biggest worry is that Pettorano sul Gizio is an outlier in terms of bear tolerance. "The other villages around, I'm not so sure. Most people live down the valley. Over there, the percentage of support for the bear decreases." When I asked what she thought were the major challenges lying ahead, she didn't hesitate. "To expand geographically. We are doing a very good job in this area, but we need to go somewhere else and follow these corridors." As the bears spill farther from their strongholds, the natural contours of the land make it clear where they will go. They will be filtered down the valleys and past the sleepy villages. Rewilding Apennines has mapped the wildlife passageways and the potential sources of conflict. It wants these routes designated as *coexistence corridors* where people and bears learn the skills to put up with each other. "When I witness bears far from the core area, that's good news," said Cipollone. "It proves the importance of the job we do." But he and Tavone are desperate to put boots on the ground before more bears arrive and do things to get themselves killed.

When his schedule allows, Cipollone jumps in his car to visit chicken coops and orchards to make sure they are "bear smart." He helps install metal doors and electric fences when he finds enclosures that are not strong enough. Large cement water tanks built by shepherds in the hills have been given ramps to allow bears to climb out safely should they fall in. Local craftspeople are hired to cover dangerous cisterns with thick metal grates. Eye-catching road signs emblazoned with "*Rallentare!*" are stenciled with the silhouettes of deer and bears to encourage drivers to slow down. Optical devices on roadside posts emit a high-pitched sound in car headlights, giving the bears a warning about traffic at night.

Like many wildlife recoveries, the work is as social as it is biological. Staff and volunteers knock on doors to get people thinking proactively about the bears. Salviamo l'Orso has a manual of best practices that members hand out in local villages. The practices include bear-proofing beehives and compost piles with electric fences and vaccinating wandering dogs against diseases transmissible to bears. Tavone has made a series of videos for landowners about how to secure their property against wildlife. The Italian villagers are more accustomed to animals than the English who live around Blean Woods. But how local people occupy the landscape still needs managing if the bears are going to continue their recovery.

* * *

As I walked the stone arches and stairways that lace Pettorano Sul Gizio a final time before I left, it struck me that the pruned apple trees are emblematic of today's carefully contrived relationship between people and many wildlife species. If animals don't have their own wildernesses to occupy anymore, then the land and its people must reshape themselves to help. This can mean creating special vulture restaurants, making bison wear shock collars, or pruning old apple trees to produce a harvest for hungry bears.

Wildlife managers back home in Montana would be appalled by the idea of pruning fruit trees to feed bears. I once bumped into a U.S. Forest Service ranger as she left an abandoned apple orchard with a backpack full of fruit. She was determined to keep the apples away from the mouths of hungry bears. Bears in wild places, the standard line goes, should be eating wild foods. None of the seven thousand varieties of domesticated apples grown around the world were cultivated to nourish them. The more bears are left alone the better.

This still might mostly be true. But in a place where bears and human agriculture have lived plow-by-jowl for centuries, these rules can flex a bit. If you are a four-hundred-pound Marsican bear and there are less than a hundred of you left on the planet, what you are eating is much less important than whether you are eating enough of it. Studies conducted in Italy in the 1970s found bears regularly supplementing their natural foods by nosing around local farms. The bears, like the griffon vultures, had come to depend on agriculture for calories. This spelled trouble when the land was abandoned by farmers moving to the cities. For a time in the 1990s, Abruzzo, Lazio, and Molise National Park partnered with World Wildlife Fund to plant corn, carrots, and apple trees on abandoned land around the park. There has also been a push for small-scale logging in Apennine forests to encourage growth of the forbs and berries eaten by the bears.

When I raised the philosophical conundrum about the appropriateness of farming the landscape for the benefit of bears, Tavone had no hesitation. "The apple trees are good for them," she said without flinching. "It is like a buffer area that gives resources to bears and helps them keep a distance from the human settlements." Rewilding Apennines and Salviamo l'Orso have together pruned over 750 fruit trees and planted 170 more. I asked Cipollone whether he thought it would be better to have bears living completely independently of humans.

"If we had pristine wilderness with no people, maybe I would be more relaxed," Cipollone told me with a smile. "It would be safer for wildlife to be wild in a wilderness. . . . But, unfortunately, we don't have it. It is almost impossible to create a buffer between humans and wildlife. It is always coexistence."

But rather than see this as a disadvantage, Cipollone saw it as an opportunity. "Some people want to have a wilderness here without the people," he said. "If you do this, I think it is the beginning of the end." Coexistence with wildlife offered something unique, he argued. "People should be proud of wildlife. It means an integral, healthy environment in which wildlife and humans can thrive. What we can do is to create this feeling of pride. Bears are not only natural but historical heritage."

The idea of bears as historical heritage was something you didn't hear often in Montana outside of Native American communities. Here in central Italy, it made perfect sense. The Italian word for *landscape* (*paesaggio*) shares roots with the word for *village* (*paese*). Both incorporate the idea of building a community. "Our central Apennines," Cipollone insisted, "should be a place where you can live and experience this connection between human communities and wildlife. The animals make our place extraordinary." People needed the wildlife, and the wildlife needed the people.

The coronavirus lockdown in Italy made this particularly clear. As the strictest part of the lockdown eased, people rushed from the towns to the mountain villages to enjoy the open space and the clean air. It was the year Amarena waltzed through the local villages with four cubs in tow. This all underscored to Cipollone the importance of his work. "It gives you the feeling you are not only saving the animals," he said, "but saving yourself in an overcrowded world."

Cipollone's description of bears as human heritage is compelling. A separation between people and animals has never made sense to the Indigenous people living alongside wildlife in other parts of the world. Once you abandon a radical separation between people and animals, the idea of one doing favors for the other becomes much more intuitive. Helping wildlife looks less like a questionable intervention and more like a reasonable condition of cohabitation. A German colleague told me how, as a child, she and her friends gathered chestnuts in the forest in the fall. They gave them to the local forester, who spread them through the winter to help the deer. In the Fens in eastern England, farmers left the tops of the sugar beet crop in the field after harvest to give geese and other waterfowl something to eat during the cold months. Today at Finland's Lake Saimaa, volunteers build snowbanks for an endangered freshwater seal so the mothers have somewhere to dig nests for birthing their pups.

The idea of animals as historical heritage does not have to draw entirely from the past. There were signs the Marsican bear was becoming increasingly woven into Abruzzo's future as its numbers increased. This being Italy, the warp and weft often involved cuisine. A local bakery offered a sweet bread, *Pan dell'orso* (bear bread). A vineyard produced a red wine called *Passo dell'orso* (step of the bear). Cipollone wanted locals to see the bear as essential to their lives. Wildlife recovery, in his eyes, was necessary for cultural recovery in this depressed area. "If we miss this, we are probably killing our future," he told me. "I don't see another way."

The glimpses he caught of villagers embracing coexistence with bears were the highpoints of his work. He told me about visiting a farmer whose chicken coop he bear-proofed a few months before. The electric fence was fiddly to set up, and Cipollone expected it to need a lot of maintenance on his return. Instead, the beaming farmer showed him an immaculately kept barrier. "I was completely astonished. I could not see a flaw," Cipollone told me, eyes wide with surprise. The farmer had adopted a whole new etiquette for living alongside wildlife.

No doubt these new etiquettes are easier to adopt when they are greased with money. The electric fences going up around beehives and chicken coops are free to villagers. Salviamo l'Orso also donates anti-parasite pills and dog food to shepherds to ease their economic burden. Sheepdogs are given microchips to help keep track of those that harass wildlife. The goodwill pays off. Cipollone told me about a blossoming friendship he has with a hunter in the village who had become an avid bear supporter. The young man had broken the traditional mold of hunters who wanted the bear dead.

* * *

I never got to see a Marsican bear during my visit, but I saw enough evidence to give me that feeling of heightened senses when walking in bear country. Tavone and Cipollone took me to several places where bears had been spotted in the last few months. They shared images from camera traps of bears walking down the same paths we walked and eating apples from the same trees we examined. They pointed to a field where they had seen a sow and cubs earlier in the spring. I saw a wooden sign with claw marks on the back. And midway through our hike, near the top of a trail in the beech forest, Cipollone stooped down in front of a large pile of dark scat. A silence settled over us as we took in the fact that a bear had recently passed this way.

Poking at the ordure with a stick, Cipollone pointed out the beech mast and the berries on which the bear had fed. After talking about the bear's diet, he used two twigs to bring a piece of scat up toward his nose. The digestive system of the bear never does a very thorough job on the berries, he explained, as he inhaled the faintly fruity aroma.

"It's like a fine wine," he said with pride before placing the scat carefully back on the ground.

OCEAN PARTNERS

1 THE WHALES WHO LEARNED HOW TO FLOSS

The parking lot at the Sitka campus of the University of Alaska Southeast may be one of the most scenic parking lots in the nation. It sits on one side of Sitka Channel looking east across the water toward the town center. Fishing boats cruise slowly through the passage on the way to deliver their catch to the local processers. Float planes and kayaks steer carefully around each other, while bald eagles and ravens wheel overhead in the salt-scented air. Across the water, St. Michael's Russian Orthodox Cathedral pokes its oxidized copper dome above the jumbled roofs of Sitka's high street. On the skyline, Mount Verstovia's snow-streaked summit hints at the jagged beauty of Baranof Island beyond. The mesmerizing scene could have absorbed me for the whole day. But I had not come to Sitka to study landscape. I had come to study whales.

At 10:00 a.m., Lauren Wild let me into the rectangular university building and took me to a small cluster of offices in its depths that house the marine scientists. When we entered her windowless workspace, her dog thumped its tail on the ground and turned a few happy circles before settling back under the desk to nap. The whiteboard behind Wild's head mapped the tasks she needed to complete before teaching resumed in the fall.

Wild is a professor of fisheries technology. Her research looks at the interactions between fishermen and marine mammals. "I study depredation," she told me. This is the phenomenon of whales preying on the fishermen's catch after they have it on their lines or in their nets. Depredation is a growing problem for fishermen all over Alaska.

Wild was born and raised in Sitka. Her husband runs a salmon boat. Her brother longlines for black cod. A study-abroad semester in Madagascar sparked a passion in her for whales. On her return from the Indian Ocean, she joined a whale project in Glacier Bay National Park under the supervision of Jan Straley, a Sitka legend in whale research. Since the mid-2000s, Wild has been working on the conflict between sperm whales and fishermen.

The problem of depredation first arose after a change in fishing practices in the region. In the mid-1990s, halibut and cod fishing moved to a quota system. Before then, fish were caught during a handful of frenzied twenty-four-hour derbies. The new quota system meant fishermen started spending much more time on the fishing grounds. The change coincided with a marked increase in sperm whale numbers across the North Pacific. The recovering whales used the newly extended season to learn how to pluck fish off the baited longline hooks used by fishermen. The valuable black cod—known as sablefish or butterfish due to their rich, silky texture—suddenly became much harder for fisherman to land. The whales were excellent theives. The scene was set for fisherman-whale relations to go into freefall. But they didn't. And the fact they didn't is a clue that big changes in how we think about whales are afoot.

* * *

Longlining is a lucrative fishery for small boat operators in Southeast Alaska. The longlines are laid on the ocean floor at the edge of the continental shelf at a depth of about six hundred meters. Hundreds of baited hooks connect to the line via a six-foot piece of nylon known as a *snood*. At each end of the longline is an anchor that keeps it attached to the ocean floor and a stretch of rope that hangs vertically from a buoy at the surface. After a six- to twelve-hour soak, the fishermen return to haul their set. They pull the buoy aboard with a boat hook, wrap the dripping line around a hydraulic winch, and start the long process of bringing the catch on board. The line gets coiled in tubs on the back deck. As the hydraulics whine, the boat pops in and out gear to keep it positioned to haul without getting tangled in the line. According to Wild, this jockeying of the engine makes a distinctive sound.

"The acoustic cue is the propeller cavitation that these boats engage in when then are hauling gear up to the surface," she told me. "That spinning of the propeller creates bubbles, and the cavitation is really loud." The whales have learned to associate this sound with food. "We have clocked them over ten or twelve miles beelining when a boat starts hauling gear. They know what it means. It's like a dinner bell."

Wild asked me to imagine it from the whale's perspective. "You have like a sushi belt coming up from the bottom," she said. "It's really hard to resist." Fisherman watch the whales feasting on their catch as the line comes to the surface. One Alaska fisherman reported a sperm whale laying alongside his boat while two more plucked fish from his line off the stern. He gave the whale a scratch behind the ears with a deck brush which the whale seemed to enjoy.

Wild spent several seasons as a fisheries researcher on boats out of Sitka, Juneau, and Dutch Harbor. She gained a deep respect for the knowledge and skills of the fishermen. Most of them, in Wild's experience, do not begrudge the whales. "They are happy to share the fish," Wild said. "They just don't want to share the fish they have caught."

Skippers told Wild about a range of different techniques employed by the whales. Some of them lunge haphazardly at whatever they see. Fish come up shredded, with teeth marks raked all over them. Others pull on the line to create tension so the fish pops off, as if understanding something about Newtonian physics. "An empty hook comes up, and you don't know if the bait fell off or if a whale grabbed the fish," Wild said. A third group engage in what the fishermen call "flossing." When a whale sees a longline being pulled onboard a boat, they grab it in their mouth. "The whale lets the main line slide through their teeth and picks fish as they come by," Wild said. The whales get the easy calories, and the fishermen have to rebait the hooks and burn precious fuel to fish somewhere else.

Wild has experimented with fishermen and fisheries managers to find solutions. "We don't want these whales to become like problem bears hanging out by the dumpsters," Wild said. She took part in a study where they experimented with several methods of gentle dissuasion.

They tried acoustic techniques. "We tested a playback device where we had recorded transient killer whale sounds," Wild said. Sperm whales are known to treat killer whales as a threat. Fishermen have observed groups of sperm whales forming a rosette with their heads toward the center as they swim away from killer whales in a defensive pinwheel. If there is a calf in the group, they will put the calf at the center for protection.

The recorded killer whale sounds did nothing. "They didn't react at all," Wild said. "I think we didn't have the right setup and maybe the wrong pod of transients."

They tried white noise and different sound combinations to get on the whales' nerves but without much luck. There are legal and ethical limits, Wild said, to how much you want to harass marine mammals with sound. They debated air bladders and flashing pieces of metal near the hooks to confuse the whales' echolocation. They discussed whether you could send a small electrical charge through the line to shock them as they tried to bite the fish. Nothing seemed viable.

Recently fishermen came up with something that appears to work. A longliner created a *slinky pot*—a collapsible, tubular pot made of fishing net stretched out by a coil of stiff wire. The pots get attached to the same longline that carries the snoods in traditional black cod fisheries. The crew cinch the pots closed when storing them on deck. When they are ready to set, they uncinch a pot, put in some bait, clip the pot to the longline, and hurl it overboard. The coiled wire opens the pot on its way to the ocean floor. Black cod swim inside to get the bait and end up trapped. They remain there until the fisherman returns to pull the gear. The whales are flummoxed.

So far, it looks promising. Nancy Behnken, a local fisherman who has used the pots to catch her black cod quota, said she was happy with the results. "There's less by-catch, and you can rebait the pots quickly," she told me over a dinner of cod collars at her house. "The whales cannot get fish out of the pots—yet!" she said. "They are smart creatures, though, so we'll see how it goes." Fishermen report the slinky pots use less bait than a line trailing thousands of hooks. You can't fit as many pots on a longline as you can baited hooks. But if you bring a higher percentage of black cod on board, it doesn't matter. The fishermen are happy, and the whales have to go catch their own fish. A good result for everyone.

It's too early to say whether every fisherman will make the switch to slinky pots. Some of them grumble about the investment in a new type of gear. But the difference between watching two-thirds of your fish going down the gullet of a sperm whale and landing everything you catch is significant. One fisherman told a trade magazine, "They've been a lifesaver for us." Wild is unsure whether the slinky pots will make the conflict with whales go away entirely. "I used to say I would like to work myself out of a job," she told me. She doubts this is going to happen any time soon. Sperm whale numbers are continuing to increase in the Pacific, and interactions with fishermen are becoming more frequent.

As I heard the story unfold, my mind spun delighted pirouettes. It was a quirky example of a natural resource conflict that required collaboration, trial and error, and lots of patience to solve. It demonstrated the high intelligence of whales as well as their opportunism. It made demands on the creativity of fishermen. Yet from the beginning of our conversation, there was something else I could not get out of my mind. It was the fact Wild had to work on this problem at all. From a fisheries point of view, surging numbers of a highly intelligent and adaptable marine mammal creates a knotty problem. But from the point of view of whale conservation, the proliferation of whales is not exactly something to lament. In fact, it's a pretty good problem to have.

* * *

Whales are the largest animals in the history of life on earth. A blue whale is bigger than the mightiest dinosaur. Its heart is the size of a small car with a beat that can be heard up to two miles away. A small child could crawl through its arteries. Whales have been known to contain parasitic worms twenty-five feet long. As well as being large, whales are also one of the most widespread of earth's creatures. They live in all the world's oceans. At their peak, they made up 90 percent of all marine mammal biomass. If you added all the seals, sea lions, porpoises, and sea otters together, you would have only a tiny fraction of what the ocean once cupped in whales.

Whales are big and also long-lived. Humpbacks average around forty-five years, but some live into their eighties. A blue whale can live over a hundred years, a lifespan revealed by counting the layers of its earwax, like growth rings in a tree. Bowheads have blue whales beat as the ocean's elders. They can live over two hundred years. One bowhead taken by Indigenous Americans in a subsistence hunt had a type of harpoon buried in its skin not used for over 120 years.

The size, mobility, and longevity of whales mean they have enormous impact on the ocean. The effect of whales on marine ecosystems is like the influence of wolves or bison on terrestrial ones. When they go from plentiful to scarce over a short period of time, the whole tangle of marine life gets yanked in a different direction.

The human relationship with whales spans millennia. Petroglyphs of whales in Australia and Southeast Alaska are thousands of years old. Shelters made of whale bones have been found at archeological sites in northern Canada and South Australia. Whales look out from face masks and totem poles carved by Haida and Tsimshian people of North America's Pacific coast. Their shapes appear in the art of coastal Indigenous people from across the planet.

The earliest evidence of whaling comes from South Korea, where etchings of whale hunts found in shale on a river bank are thought to be eight thousand years old. Indigenous people sharpened antlers, shells, and rocks to use as tips of their whale harpoons. Whaling was dangerous and only occasionally successful. But the immense size of a whale meant the payoff was huge.

The Basques commercialized whaling in the North Atlantic a thousand years ago. Towns on Spain's Atlantic coast still feature whales in their coats of arms. Between 1530 and 1610, Basque whalers killed more than forty thousand right whales off the coasts of France and Spain. As they depleted the local populations, they were forced to whale farther and farther from home. By the seventeenth century, the Basques were killing bowhead whales off Spitzbergen. By the eighteenth, they were hauling whales out of Baffin Bay. Right whales and bowheads were soon almost gone from the North Atlantic.

Whales had numerous commercial uses. Their flesh provided calories, but it was their oil where the money lay. The average humpback whale held thirty-seven gallons of oil. Whale oil rendered from slabs of blubber went into lanterns, candles, and industrial soaps. It was used in insecticides and served as a lubricant for the cogs of factory machinery. Printers put it in their inks, and artists used it to keep the colors in their palettes smooth. Tanners valued whale oil for how it helped them soften their leather. In the late 1800s, whale oil was used in the manufacture of nitroglycerine. Tractor operators poured it into their hydraulic systems. The American car company General Motors put it in transmission fluid until 1973.

A particularly valuable oil, known as *spermaceti*, was lifted from the skull cavities of sperm whales by the bucket. It commanded a high price for cosmetics and ointments. Ambergris, a waxy substance produced in a sperm whale's bile duct, was used as medicine and a fixative in perfumes. A hundred-pound lump could sell for $20,000 in the late 1800s. For a long time, the origins of ambergris were shrouded in mystery. Twelfth-century Chinese herbalists claimed it was dried dragon spittle. The dull, gray nuggets sell for as much as platinum in countries where the trade is still legal.

Whale oil became for a time a miracle cure. A medical hotel specializing in whale therapies sprang up in Australia. Whale bodies became temporary sarcophagi so the affluent could expose their skin to the potent juices and odors. Australian author Rebecca Giggs quotes from a *New York Times* story published in 1896: "The whalers dig a sort of narrow grave in the body, and in this the patient lies for two hours, as in a Turkish bath, the decomposing blubber of the whale closing around his body and acting as a huge poultice." In World War I, whale oils were used as a salve to stave off trench foot. The oil also found favor in more tranquil environments as a high-end massage oil.

The other commercial driver of whaling alongside oil was baleen. Baleen comes in plates up to ten feet long attached to a whale's upper jaws. It is made of keratin, the same material found in human fingernails. The whale uses the bristles on the plates' lower edge as a filter. There are more than a dozen types of baleen whales, including humpbacks, bowheads, right whales, gray whales, and giant blue whales. A humpback can have as many as eight hundred baleen plates in its mouth. The whale takes in huge mouthfuls of sea water and then expels the water through the baleen by pressing their tongue against the roof of their mouth. The sieve-like baleen keeps fish and krill trapped after all the water has drained away, like gnocchi on a slotted spoon. Baleen whales tend to eat large quantities of relatively small prey. A humpback's throat is not much bigger than a large grapefruit, so it will spit out any seabirds, seals, or human divers that accidentally end up in its cavernous maw.

Indigenous people harvested whales for sustenance and used the baleen to make baskets and roofing. Commercial whalers found a much wider range of uses for the strong and flexible product. It provided stiffness for buggy whips, shoe horns, and fishing rods. It went into umbrellas, toothbrushes, hat rims, pastry crimpers, tongue scrapers, back scratchers, throat swabs, and surgical stitching. Toymakers put it in hula-hoops and fashioned it into springs to power wind-up toys. Surgeons used baleen to build prosthetic limbs. Upholsterers used it as sofa stuffing. There were few rooms in an affluent, early twentieth-century home that did not contain some part of a whale.

Whales gave shape to society, sometimes literally. Giggs points out that the hourglass ideal imposed on a woman's body is a legacy of whaling, thanks to baleen's role in stiffening corsets and skirt hoops. Police wielded it as nightsticks, and boarding school headmasters thrashed children with baleen canes. Like bison slaughtered on the great plains, when all other uses were done, whale bones were ground up and spread on fields as fertilizer.

Nineteenth-century whaling took a significant toll on whale populations. The United States employed seventy thousand workers in the whaling industry at its peak, one in six of them African American. Yankee whalers wreaked havoc on land and at sea. Between whaling trips, they ate Galapagos tortoises and spread invasive rats to remote islands around the world. When the U.S. plunged into the chaos of its Civil War, British and Norwegian whaling fleets picked up the slack. Sandefjord on Norway's southern coast contained a third of the world's whaling companies at the start of the twentieth century. The large villas overlooking the ocean today are a legacy of a time when it was the richest city in Norway.

Fossil fuels allowed whaling to spread to the remote Southern Ocean in the twentieth century. Whale species previously hard to catch—like blue, fin, and sei—became the targets of larger and faster boats. Compressors pumped bodies of slaughtered whales full of air so they would not sink. Plummeting whale populations in the North Pacific and Atlantic were soon followed by similar crashes in the Antarctic. Humpback whales off South Georgia were first caught in 1904. They were close to extinct by 1915. In some years, more than three thousand blue whales were killed in the cold southern waters. Factory processers stayed at sea for months on end, servicing a fleet of nimbler hunting vessels. Whales no longer had a refuge.

In the mid-1910s, a new market for whale oil emerged after the development of hydrogenation. Hydrogenation was a way to solidify the oil and take away its smell, making it suitable for use in products like margarine. Not long afterward, the blue whale unit (BWU) was introduced to quantify whale harvests. A BWU was the amount of oil that could be extracted from one blue whale. It was deemed equivalent to the oil in two fin whales, two and a half humpbacks, and six sei whales. The whales' transformation into a market commodity was complete.

Whaling should have faded in the twentieth century. Kerosene and electricity were better sources of light. Synthetic oils and plastics made the use of whale oil and baleen increasingly unnecessary. Instead of decreasing, however, whaling surged. The Soviet Union and Japan joined the fray. At least three million whales were killed in the twentieth century.

Cold War politics made global agreements to stop the whales' decline difficult. Blue whales in the Antarctic collapsed from nearly a quarter of a million to fewer than five hundred. Humpbacks in the North Pacific dropped to twelve hundred. In the western South Atlantic, they declined to 450. The commercial industry had wealth and political power. Public sympathy toward whales barely existed. Plus, the margarine still tasted good on toast.

By midcentury, it was evident to everyone that whale populations were being wiped out. Tribes had long ceased subsistence hunting because whales were so rare. The International Whaling Commission was founded in 1949 to regulate the industry, but it failed dismally. Quotas set in BWUs encouraged whalers to go after the bigger species. Whaling companies rushed to maximize their share of what was left.

By the time the IWC banned the killing of humpback whales in 1966—following it up with a moratorium on all commercial whaling in 1986—the whaling industry had more or less run out of whales to catch. The Antarctic populations had gone the same way as the North Pacific and Atlantic stocks. Millions of whales had been killed since the Basques threw their first harpoon. The number of great whales crashed by 90 percent, with right whales and blue whales pushed to the brink of extinction. At least 300,000 humpback whales were killed. Southern Ocean sperm whales were reduced by 95 percent. Eighty-five percent of the biomass of whales in the Antarctic had been plucked from the sea and taken to far-off continents in the holds of whaling ships.

The whaling moratorium imposed by the IWC hardly covered anyone in glory. It was decades overdue, and it was driven by commercial, not ethical, imperatives. But the moratorium did finally provide whales with just a sliver of a chance. And the way whales seized that chance is one of the most instructive stories in conservation.

* * *

Risø is not the kind of coffee shop you expect to find two hundred miles north of the Arctic Circle. The baristas wear pressed white aprons and sport elegantly waxed moustaches. Jazz music softens the corners of a room that burbles with conversation and the clinking of silverware. Old friends and college students sit crowded around small tables, hunched over plates of eggs and Norwegian salmon. The weak November light paints everything amethyst outside.

I arrived early to enjoy the atmosphere, knowing my meeting at Risø would likely be my only outing that day. The arctic winter encourages long hours cozied up at home. The Scandinavians are experts at *hygge*, the art of feeling warmth in simple things. The foam on my coffee drink had been scribed with a hovering ghost, a nod to the recent Halloween holiday. I took a sip and waited for the scientist who had promised to share with me the basics of whale recovery.

Martin Biuw showed up wearing a woolen Norwegian sweater and a huge grin. His face was roughened by a graying stubble. He looked like a smaller version of Viggo Mortensen on a break from the movie set. Biuw is Swedish by birth. He and his wife share a two-story home a few miles out of town on Kaldfjord. Through a gap in the mountains, they can see the steep peaks of neighboring Ersfjord, the setting for thousands of postcards of Tromsø's famous northern lights.

Biuw is a senior scientist at the Institute of Marine Research in Tromsø. He spends weeks aboard heaving research vessels on remote oceans and long winter months crunching data from field sites around the world. His work on marine mammals has bounced him back and forth between the Arctic and Antarctic Oceans for more than a decade. It is a piece of sublime luck that also he happens to live on Kaldfjord.

The scenic fjord, located ten miles outside Tromsø, caught the attention of Norway's marine scientists a few years ago after scores of humpback and killer whales arrived to feed on gigantic schools of herring. The herring were not a total surprise. Biuw told me he had looked at church records spanning the twentieth century and found accounts of occasional superabundances of herring.

"They mention these peaks of herring in Kaldfjord," Biuw said, "but they don't say anything about whales." The humpbacks had been too depleted. The scarcity of whales had not always been the case. Biuw had talked with an elderly lady in Kaldfjord who said her grandparents told stories of large numbers of whales pursuing the herring into the fjord in the 1800s. "Probably, what we see now," Biuw said, "is what we used to see before the whales were hunted." Kvaløya, the island containing the fjord, translates into English as "whale island."

After whales started pouring into Kaldfjord in 2011, an instant whale-watching industry sprang up to cash in on the surprise visitors. Europeans already flock to Tromsø each winter for a glimpse of the northern lights. If it is cloudy, which it often is, they entertain themselves with sled dog rides or by eating reindeer stew with herders in a *gamme*, a type of traditional Sami hut. These carefully curated experiences often end around a fire with some traditional Sami singing, known as *joiking*. For those determined to see the northern lights, Tesla EVs drive guests a hundred kilometers inland where the weather tends to be more cooperative.

But now they had whales, and a fleet of boats quickly assembled to take advantage of the bonanza. I went out on one—an oak-hulled schooner converted from a previous life as a fishing trawler—that had come down from Iceland for the whale-watching season in Kaldfjord. *Opal* was a gorgeous broad-beamed sailboat with two masts, a bowsprit, and a tangle of rigging. The wooden deck had plenty of vantage points from which to scan the fjord for whales. The twenty tourists on board were stuffed into bulky mustang suits that made them look like astronauts on a spacewalk. A wood stove crackled below decks. Hot soup was available from the galley when it was time to warm up.

Standing outside at the wheel in the stern, the skipper, a bearded Icelander, told me he takes the boat to Greenland during the summer. He was young but oozed a kind of Viking competence that said we weren't ending up at the bottom of the fjord under his watch. Winter whale watching in Norway was relatively new for him and a welcome addition to the calendar. The *Opal*'s silent electric motors gave them an advantage over other whale-watching vessels in Kaldfjord's darkness. As we glided up the fjord through frigid, arctic air, the skipper talked global politics and fishing. He chided the Norwegians for still hunting whales, something he claimed Icelanders, despite officially still being a whaling nation according to the IWC, had more or less left behind.

It wasn't our night for humpbacks, but the excitement remained high. The northern lights appeared briefly, and the skipper pointed out a killer whale's dorsal fin breaking the surface in the darkness. Frequent trips below decks to warm up by the fire or grab some soup provided plenty of opportunity to listen to crew members wax about the sublimity of summers spent on the Greenland coast. The schooner was experiencing a second life in northern waters in the tourist economy, a life that could last much longer than the first.

Biuw pointed out whale watching was not the only economic boom in Kaldfjord. The herring schools were great for local fishermen. Unlike the big boats that could follow herring wherever they spawned, the smaller, local operations had only enough equipment and budget to fish close to shore. When the herring filled up Kaldfjord, it created a once-in-a-generation opportunity. "The entire coastal fleet came to the Klondike of herring fisheries," Biuw said.

Like all booms, this one was temporary. After five years, the herring moved on, for reasons not entirely clear. Both the herring and the whales are still around, taking advantage of fjords farther to the north. The fact humpbacks came to Kaldfjord in such numbers at all was a reminder to everyone of how well they are doing. It had been a century since Kaldfjord was awash in so many whales. It delighted tourists, small children, fishermen, and scientists in equal measure.

Biuw didn't hesitate when I asked him which whale species had made the most progress. "The humpbacks have made the most dramatic recovery," he said. "But also the fin whales." Fin whales are named for a sharp ridge that stretches between their dorsal fin and their tail. They are less well known to whale watchers than humpbacks because they commonly feed farther offshore. But fin whales are fast. "They go like a projectile through the water," Biuw said. "It's very dramatic." When hunting, they turn on their side and lunge at fish or krill on the surface with their mouths partially open. Biuw told me of a recent expedition to the Antarctic to measure krill abundance. He was expecting a healthy humpback population, but he was astonished by the glut of whales he saw.

"When we came close to the South Orkney Islands, we started seeing a huge number of birds, albatrosses, and all kinds of things, and we realized something dramatic was going on. Then we started seeing whale blows all over the place. There must have been fifty or sixty fin whales all feeding on this large school of krill. There were three or four blue whales mixed in. It was one of these feeding frenzies. Albatrosses everywhere. It was absolutely spectacular."

Despite a few distinct populations remaining vulnerable, both fins and humpbacks across northern and southern oceans have rebounded. Some of the increases involve breathtaking numbers. Humpbacks in the western Indian Ocean have grown from around six hundred to 36,000. Populations off both east and west coasts of Australia passed 50 percent of their preexploitation levels in 2015, with growth rates of more than 10 percent a year. The number of humpbacks in Glacier Bay, Alaska, went from forty-one in 1985 to 239 in 2013. Worldwide, the species may be completely recovered by the end of the decade.

Sperm whales in the Pacific also appear to be increasing. Southern Ocean right whales had recovered from a low of three hundred in the 1920s to over fifteen thousand today, with some populations growing at 7 percent a year. Minke whales are categorized by the IUCN as a species of "least concern" and are still being hunted by Norway and Japan. Bowhead whales in the Bering, Chukchi, and Beaufort seas have tripled and are close to preexploitation levels. Passengers on one vessel recently sailing off the Antarctic Peninsula were treated to an experience even more remarkable than Biuw's. They saw over a thousand of the typically solitary fin whales congregating in one superpod.

All of these bits of good news should be put in perspective. Many whale populations still struggle. North Atlantic and Pacific right whales stand on the brink and may be headed for extinction. Bowhead whales around Svalbard and the Okhotsk Sea number in the low hundreds. Blue whales, the ocean's mightiest whale, are definitely not out of the woods. They remain critically endangered in Antarctica, where their population is only 1 percent of historic levels. Gray whales have been dying all along the Pacific Coast in worrying numbers. Threats to whales from pollution, fishing gear, ship strikes, and climate change are increasing every year. The noise caused by boat traffic and wind turbine installation is beating down the edges of the whale's acoustic world. Short-lived warming events like "The Blob," which took place in the Pacific from 2013 to 2016, have dire effects on feeding and reproduction. In 2020, more than 350 experts from around the world signed an open letter to global leaders warning that half of cetacean species are still under threat. Nobody should be under the illusion that whales are out of danger.

Yet the whales that are doing well are important. Not only do they show that *not killing* an endangered animal can be a remarkably effective conservation strategy; they also open doors to new ways of thinking about wildlife. Whales were once regarded as an exploitable resource that could be killed freely for profit. No more. Millions of people now value them as a watchable treasure worth protecting. Many whale watchers also value them inherently for what they are in themselves, whether or not they ever get the chance to see one. A whale's size, charisma, and intelligence make it worthy of protection in its own right. This is clearly an ethical advance, a small sign of moral maturation in our species.

I was, however, about to learn an even more interesting ethical perspective on whales, something deeper than just giving them a higher rank in value. Whales have important lessons to teach about how the lives of recovering animals can fit helpfully alongside our own. To understand what these lessons are, I needed to take a closer look at the resurgence of humpback whales in the North Pacific. I also needed a different kind of scientist to unlock that ethical door.

2 A GREAT STIRRING

The large windows in Heidi Pearson's office look out on the steel-gray waters of Alaska's Auke Bay. Bald eagles twitter from the branches of nearby spruce and hemlock trees. Assorted boats trail frothy V's as they head out from the harbor into the channel below. In the middle distance sits Coghlan Island, its soft contours a teaser for the snow-capped peaks of the Chilkat Range beyond. If you wanted a snapshot of Southeast Alaska in midsummer, this is it.

Pearson welcomed me into her office and admitted how sweet it was to spend her workdays here. She can glimpse the misty exhalations of whales from her desk and watch the weather create changing patterns on the ocean's surface. The newly renovated Anderson Building is some of the best real estate on the University of Alaska Southeast's Juneau campus—doubly so for someone in Pearson's line of work.

Pearson is a marine mammal behavioral ecologist. Simply put, this means she looks at how creatures like whales, dolphins, and sea otters fill their day. She grew up in Iowa, more than a thousand miles from the nearest ocean. Her interest in the sea took off after getting certified in scuba in high school. One memorable spring break, her class boarded yellow school buses to Florida and spent two weeks camping on the beach.

They snorkeled and dove and spent an afternoon swimming at Crystal River with the manatees. Pearson was hooked.

In her early days, Pearson divided her loyalties between the ocean and the land. "When I was fifteen, I went to East Africa," she said. Her parents worked in education. One of their colleagues took a group of talented students, including Pearson, to see mountain gorillas in the country then known as Zaire. The trip was a game changer. "We hiked through the rain forest for a couple of hours and came up on one of the habituated groups," she said. "We spent about an hour with them." She had never been so close to a wild primate. "During that hour, we were all just hanging back, and there was a small juvenile that came and walked right across my feet." It felt personal. "Just to be in this other species' environment," she recalled, "and to have them so trusting of us." She knew her life's work would be in conservation.

Pearson went back to Africa to study colobus monkeys while pursuing her undergraduate degree at Duke University before turning her attention to marine species. She enrolled in a Ph.D. program in wildlife and fisheries at Texas A&M specializing in marine biology. The first field opportunity she got was to help out with a sea otter study in Alaska's Prince William Sound.

For two summers, Pearson lived at a rugged backcountry camp in Simpson Bay, documenting the territorial behavior of male sea otters. She sat for hours in a skiff with a long-lens camera and notebook. Sea otters spend more time at the surface than any other marine mammal, making them a good study subject for people in boats. Her stopwatch beeped once a minute, which was her cue to note what the otters were doing. The field crew mapped the otter's habits and movements, building a picture of how the males defended their territory. In addition to the territorial findings, they published a paper on how to identify sea otters by the distinctive marks on their noses. Nose scars collected from fights and the sharp edges of clams create something akin to a fingerprint unique to each animal.

Pearson left the sea otter project to focus her dissertation on dusky dolphins. Since her time with the mountain gorillas, she had been fascinated by the question of what large mammals gain from associating with each other. "Dusky dolphins have these complex fission-fusion societies," she told me. "This is what human societies are too. We are constantly forming different groups with different people, forming different relationships. Some people we have stronger bonds with than others." She wanted to work out why animals bother building this community at all. In one section of her dissertation, she compared what she learned from dolphins to what is known about primates.

"I'm interested in how these social relationships are formed in these very different animal societies that have evolved in the water versus on land." One hypothesis for the presence of large brains in primates and cetaceans is they evolved to manage their complex social lives. "You need a large brain with a lot of processing power to keep track of these relationships," Pearson said. "Who is your friend? Who is your enemy? Who do you want to form an alliance with?" One thing was clear right away, a philosophical point that would increasingly haunt me the more I learned about whales: in Pearson's line of work, the lines between humans and the rest of the biological world are not always firm. When she investigates the lives of marine mammals from a boat, she finds constant reflections of our own.

* * *

One of the reasons I had come to Alaska was to see Pearson at work on the water. As a behavioral ecologist, Pearson knows more than most people about how to be in the company of whales. Humpback whales return every year to Southeast Alaska after spending the winter in calving grounds off Mexico and Hawaii. The journey is an epic feat of navigation and endurance. Humpbacks have one of the longest migrations of any mammal, swimming farther than the average American drives in a year. The females arrive in their Alaskan feeding grounds in late spring, taking about forty days to travel from their calving grounds. Some of them show up with a new calf in tow. The males join later in the summer, with most of them back near Juneau by early July.

Migrating humpbacks, like salmon, have high site fidelity. This means they return to the same spots each year. Some whales have been tracked feeding in the same bays in Southeast Alaska for forty years. Pearson has spent the last decade building a detailed record of the whales that visit her study site, getting out on the water most weekends if she can. She has documented more than 170 different humpbacks around Juneau in summer. At this point, she knows quite a few of them by sight.

"We will probably see Riddler tomorrow," she told me the day before we went out on the water. Pearson hoped we might also see a mother and calf pair recently returned to the feeding grounds after their 3,500-mile journey. Waiting for their arrival is like anticipating the return of old friends from a six-month tropical vacation. They have traveled far, but they're the same old friends she knows so well. If she spots a whale she cannot identify, Pearson feeds the images into the website HappyWhale.com, which digitally compares the pictures to a database of West Coast humpbacks.

On a cloudy Saturday morning in mid-July, I joined Pearson, her husband, Chris, and their two-year-old daughter, Maia, in their twenty-one-foot Boston Whaler for a humpback photo ID survey. As we left Amalga Harbor, we motored slowly past two seine boats with their nets laid out. Pink salmon jumped out the water not far from the giant C's carved by lines of floats on the rippled surface. After we passed the seiners, Chris gunned the 250-horsepower outboard, and the boat picked up speed, lifting itself free from the water's clutch. With Maia bouncing on her parents' knees, we set out into North Pass to find a whale.

The main tools you need to identify humpback whales are a rapid-fire camera with a long lens and a reference list of whales. Every humpback tail (or fluke) has a unique pattern of colors, scrapes, and indentations. The flukes look like fat and blotchy airplane wings. Getting a clear picture of these patterns is the key to identifying which whale you are looking at.

The survey process begins when you spot, or hear, a whale blow. A small jet of mist will explode from the ocean surface accompanied by a sound like a compressor discharging air. The mist sprays ten feet straight up and hangs there for a few seconds before slowly dissipating. It is visible more than a mile away. A humpback will usually take four or five of these recharging breaths before making a sounding dive toward the bottom to feed.

To make the sounding dive, the whale has to jackknife its body from horizontal to vertical. As it arcs, the whale's dorsal fin breaks the surface before a body part known as the *caudal peduncle*—twenty-feet of tapered muscle between the midriff and tail—rises out of the water. The peduncle curves at a tightening angle, rising higher and higher from the water as the whale heads for the depths. Finally, the flukes break free and fly briefly above the surface like a farewell flag.

The whole thing looks like it happens in slow motion, but the flukes leave the water for only a couple of seconds. This is the moment you need to be behind the whale with your camera ready. The underside of the tail—visible only from behind—has a much more distinctive pattern than the topside. It is the key to successful whale identification. "In the last couple of summers, I have got a lot better at knowing them from their dorsal fins," Pearson said. But reliable identification always requires a clear photo of the bottom of the flukes.

The first thing Chris did when we arrived in the survey area was to stick a green flag that said "Research" onto one of the vessel's rails. This was to let other boats know we had a permit to photograph whales up close. Pearson was now on home ground. She skipped around the boat readying herself for the day's work with obvious joy. Some of the whales propelling themselves through the waters beneath us had first shown up on her camera more than a decade ago. Pearson wanted to know how they were doing. We didn't have to wait long before a humpback surfaced not far from the shore.

Pearson said almost immediately, "I think that's Eyelet." Chris turned the boat in the whale's direction and picked up speed. Maia grinned, her face surrounded by the blue halo of her life vest. "Another whale just surfaced at 12 o'clock," Chris said as he drove. Having skippered a whale-watching boat a previous summer, Chris was always scanning for the next target. "Let's get this first one and then head there afterward," said Pearson.

"How do you know that's Eyelet?" I asked over the engine noise as we closed in on where the whale had broken the surface. We had not seen the flukes, and the whale had shown part of its back for only a few seconds.

"I think I recognized the dorsal," Pearson said, "and I have seen Eyelet hanging out in this area before." I forgot. She knew these individuals well.

When looking for food after a sounding dive, humpbacks usually stay down for about eight minutes before surfacing again. Pearson set her stopwatch, and we swiveled our eyes in all directions, not quite sure where the whale would come up. More or less on schedule, it resurfaced farther along the coast, and Chris accelerated to position the boat a comfortable distance behind where it rose. We could see a smooth patch of water known as a *flukeprint* still glistening on the surface that marked where the whale had submerged. The whale broke the surface again, and Pearson's camera clicked rapidly, getting a good image of the flukes. She pulled up the catalog on her field laptop and showed me how the photo on her camera matched the image of Eyelet in the catalog. After making some notes about the whale's position and behavior on the data sheet, we headed toward where Chris had spotted the next whale.

"If that's Nibblet, she's parked her calf," Pearson said as we closed in on the area. Keeping track of mother-calf relationships is part of Pearson's study. The previous summer, she joined marine biologists from the National Oceanographic and Atmospheric Administration and from the University of Alaska in Fairbanks to collect hormone samples from several whales. They took tiny plugs of blubber with a dart gun and flew a drone dangling a petri dish into clouds of warm whale breath. The samples were sent to labs in Juneau and West Lafayette, Indiana, for analysis. The results would provide a baseline of stress levels for local humpbacks.

During the tests, they discovered at least five of the females were pregnant. Three of them were already back in the study area, though Pearson was yet to confirm they all had calves. A fourth, the pair she was hoping we might see today, had been spotted ninety miles south in Frederick Sound. Usually, Pearson would see a mother and calf swimming together, but the females also took off on feeding forays alone. They left the calves parked in a bay, returning to pick them up later. There were a few killer whales in the area but not enough to frighten the humpbacks. I asked if the whales ever assigned babysitters. Pearson said there was no evidence for it. The humpbacks were confident enough the killer whales would stick to seals and sea lions.

For the next several hours, we never waited long before seeing the fine mist of a whale's breath erupt from the ocean surface. One whale, named Sasha, had marks on her skin from a run-in with some fishing gear. Riddler showed up as expected with his distinctive notched dorsal. Juneauite was a very quick ID, cruising her regular territory with a distinctive fluke. Pearson told me they were worried about Juneauite. Her shoulder blades jutted out, something you would not expect if she were a healthy weight. Flame appeared with her calf parked somewhere unknown. It was when Flame swam directly toward the boat and arched her back for a sounding dive that I got a sense of how massive a humpback is.

"Yes, very girthy!" Pearson remarked after seeing my silent "Wow!"

We saw ten whales over the next several hours. I was astonished at how many were feeding in this one area. We had made a haphazard circle around Shelter Island. The circle was less than ten miles across. Pearson informed me this was a decent day's spotting but not an exceptional one. There were days when she saw more. Today's whales, she said, had been mellow. No breaching. No lunge feeding at the surface. But the sheer number was impressive. This was one island in one inlet off one portion of Southeast Alaska's coast. If you multiplied this by all the bays and islands across the region, it added up to a whole mess of whales—thousands and thousands of whales stirring the ocean with their flukes as they wove three-dimensional patterns through the depths.

Back in Amalga Harbor, we tied up to the dock to wait our turn on the boat ramp. Chris went to get the trailer while Pearson started closing up the computer and putting away her equipment. Although Pearson's primary interest is in whale behavior, her study has made her think more and more about what this resurgence of whales means for the ecosystem. From a low of around twelve hundred when the hunting moratorium took effect, humpback numbers in the North Pacific have ballooned to over twenty thousand, a figure believed close to preexploitation levels.

After half a century of absence, some of the world's largest mammals are once again living, feeding, and breathing their last in the cold waters of the North Pacific. If whales were once 90 percent of marine mammal biomass, their loss and recent recovery must be having an effect on the surrounding system. This raised some intriguing questions. How much are recovering whales reshaping the ocean environment? And what does their recovery mean for fish, for phytoplankton, and, ultimately, for us?

Pearson has recently embarked on a new mission to find out.

* * *

In 2015, a conservation biologist at the University of Vermont named Joe Roman coauthored a paper with a group of colleagues from the UK, Denmark, and the Netherlands titled "Global Nutrient Transport in a World of Giants." The paper's thesis is that large animals previously played a measurable role in transporting nutrients across ecosystems prior to their near-extinction in the late Pleistocene and Holocene. A constant supply of nutrients is essential for the living world to function. Without them, life's profusion would slow to a trickle before ultimately snuffing out.

The dominant view of scientists is that wind, fire, and water do most of the heavy lifting when it comes to nutrient transport. After being released from rocks by natural weathering, nutrients are washed and blown in various directions by climatic and geological forces. They spend a few years in the bodies of animals and plants, but those stopovers don't alter the fundamental dynamics of the system. It is rainfall and gravity that move elements such as phosphorous, nitrogen, and iron through the system toward their final resting places at the bottom of the world's oceans.

Roman and his fellow authors challenged the established wisdom. They proposed that sizable quantities of nutrients move in the bodies of creatures that walk, swim, and fly them onto different parts of the landscape. This was especially true in the past. "The past was a world of giants," they wrote, "with abundant whales in the oceans and terrestrial ecosystems teeming with large animals." These giants contributed to nutrient cycling through their wanderings and migrations. Whales, woolly mammoths, and steppe bison all moved massive amounts of earthy goodness against the pull of gravity before depositing it in their waste. When the megafauna went extinct, these movements slowed.

Today the problem is not just diminished numbers of animals. The remaining wildlife are squeezed into tighter home ranges by agriculture and development. The domestic animals that have replaced them don't do much to help. Vast numbers of pigs and cattle in industrial feed lots eat and deposit their waste in the same few square feet. The authors estimate that on every continent except Africa, nutrient movement by animals is less than 5 percent of the past.

As some megafauna recover, Roman is convinced it is time to revisit the science. "We must wonder," he and his coauthors asked, "What role do animals play in transporting nutrients laterally across ecosystems on land, vertically through the ocean, or across the ocean-land divide?" They note how wildlife can move nutrients in particularly interesting ways. Moose dunk their massive heads to graze the bottom of ponds and then move those nutrients onto the land. Seabirds wing them from ocean to shore by eating at sea and nesting on cliffs and beaches. Migratory fish such as salmon and striped bass ship nutrients inland when they leave the ocean and ascend a river to spawn. Bears, mink, and vultures take the nutrients packaged in these fish and move them out of the streams and into the forest. Wildlife use their muscles to lift nutrients across boundaries that gravity would never permit. They heave the constituents of life from place to place before gifting them to their surroundings when they defecate or die.

The most generous animals in this planetary gift economy could well be the great whales. One reason for this is their size. The larger the animal, the more nutrients they shift. Blue whales in the North Pacific consume up to sixteen tons of krill a day. This adds up to masses of goodness warehoused in a blue whale's body. Roman and his collaborators calculated that a hundred-ton lactating blue whale ejects 3.3 kilograms of nitrogen out of her body on a daily basis. Nitrogen is key for constructing the proteins all of life needs. Iron is also important, essential for oxygen transport in living organisms. Iron concentration in whale fecal plumes is ten million times stronger than ambient levels of iron in the Southern Ocean.

Whales play an especially important role in nutrient transport when they feed in one part of the ocean and fast in another. Humpback whales eat two tons of krill, herring, and sand lance a day in productive waters. Gray whales consume over 300,000 pounds of marine organisms in four months on their northern feeding grounds. The humpbacks, blues, and gray whales then leave their feeding grounds, swim up to five thousand miles, and spend several months in warmer waters breeding, giving birth, resting, and singing. They eat very little during this time, losing up to a third of their body weight. Much of the lost weight is emitted as feces and urea (which go into the water) or as milk (which goes into the bodies of their calves). It is a remarkable feat of biochemistry. It's also a pretty impressive logistical accomplishment. The fattened bodies of whales act as watertight shipping containers as they transport nutrients from polar to equatorial seas.

The movement of nutrients by whales from high Arctic and Antarctic latitudes to low equatorial ones is known as the "great whale conveyor belt." In past centuries, it was a mighty machine. Baleen whales alone ate twice the weight in krill of all today's fisheries combined. Roman has estimated that blue whales might have moved 24,000 kilograms of nitrogen from the Southern Ocean to warmer waters before whaling. Nutrient movement by whales from Antarctica has declined by at least 98 percent since then.

The role of whales in nutrient cycling does not end with this trans-oceanic conveyer. They also move nutrients from the bottom of the water column to the top. In every ocean, gravity is constantly pulling nutrients down below the photic zone (the part of the sea penetrated by surface light). They fall as *marine snow* into the darker benthic zone. The snow is made up of fecal matter, dust, and the bodies of diatoms, bacteria, larvae, and small crustaceans. After the snow has dropped below the photic zone, it stays near the bottom. Cold temperatures and a lack of mixing by wind and waves prevent it from moving back up. This is a limiting constraint on ocean life because the photic zone is where all the photosynthesis occurs. This limit is a big deal. A healthy photic zone is responsible for half the photosynthetic activity on the planet.

Whales have the capacity to swim nutrients up through the falling snow. By eating at depth and excreting at the surface, they kick-start the food chain by bringing nutrients essential for photosynthesis to the surface. This vertical movement is known as the *whale pump*. The authors of the "World of Giants" paper calculated marine mammals may have moved 340 million kilograms of phosphorous from depth to the surface before whaling began. They estimate nutrient movement is down by nearly four-fifths thanks to the reduction in whale numbers. They think the recovery of whales can help restore both the conveyor and the pump.

When I asked Biuw, the scientist in Tromsø, about the effects of whales on nutrients, he was open to the idea but noncommittal. "There have been some really interesting papers written about what the role of large whales might have been for ocean fertilization," he said. (I learned later he was talking about Roman's work.) Biuw was confident whales had an impact on ocean ecology but was not willing to speculate on the extent to which they act as global conveyors or vertical pumps. "There is quite a lot of controversy about it," he said. "There is some serious scientific criticism from people saying that the whale population is too small."

But then he added that the waters in some systems, such as the Southern Ocean, are relatively confined by ocean currents. The Antarctic Circumpolar Current forms a vortex of moving water around Antarctica. The vortex acts as a barrier to ocean mixing. Biuw could imagine a whale pump having an effect there. "It could be that all the nutrients are retained within that system, and then it might have a locally strong effect," he said. When I got home, he sent me a slew of Roman's papers.

Pearson is also careful to point out that the science is not yet solid enough to conclude recovering whales are boosting the ocean's productivity. She thinks the hypothesis is promising but knows there are gaps in the research. "We can probably safely say that they must be having *some* impact," she says. "But is it a biologically meaningful impact?" She does not want to overstate the case before the data is in.

Fortunately, data is exactly what Pearson is in the business of collecting.

* * *

"This looks scarier than it is."

Melissa Rhodes-Reese had noticed my baffled expression as we gazed at the lab instrument sitting on the bench in front of us. The instrument was a black metal box about the size of a carry-on suitcase. It had a half dozen moving rotors at one end, a smattering of red and green LED lights, and dozens of thin plastic tubes sprouting in every direction.

"It's all plumbing," Reese added, trying to be reassuring.

The two of us were standing in a windowless room deep within the University of Alaska Southeast's environmental science building. The bland, gray and blue structure sits on a gravel lot at the northwest end of Gastineau Channel, a good mile or so away from the rest of campus. Jets on their final approach to Juneau Airport whine overhead. Reese called the building "the ugly stepchild of the university," an observation not entirely out of order given the otherwise scenic campus. In addition to the lab in which we stood, the building also housed chemists, hydrologists, and foresters, all probing the secrets of this rare coastal ecosystem. Below us, scientists from the Alaska Department of Fish and Game rented space in which they cut up hatchery salmon to extract coded wire tags and otoliths (ear bones). What the building lacked in aesthetics, it made up in raw science power.

The machine Reese and I were examining was a nutrient analyzer. In one end go liquid samples of what you want to analyze. Out the other comes a computer readout of nitrate, nitrite, ammonia, and other nutrient levels. The machine looks complicated, but the target chemicals are run-of-the-mill. What is not so common is the material Reese is analyzing. She and Pearson are conducting a study to look at the nutrient levels of whale poop.

The first challenge they faced was to figure out how to run such a study. "There is no tried-and-true whale-poop-analysis method right now," Reese told me. Scratching around for ideas, she called up an aunt who is a soil scientist at South Dakota State University. "You are in ag school," she said to her aunt. "You know how to analyze poop. What can we do?"

Her aunt had an analyzer similar to the one sitting on the table in front of us. She told Reese to come to South Dakota with a few samples. Reese flew to Brookings with a cooler, creating, she told me, a few giggles among the TSA agents. Her aunt showed her how to run an analysis similar to how the agriculture people look at fertilizer and cow patties. They mixed the samples Reese brought with liquid reagents and let the machine pump the resulting brew through various tubes. A sophisticated optical device read the color of the liquid. The wavelength of the light allows a computer to plot its chemical signature. Back home in Juneau, Pearson and Reese wrote a grant to buy their own machine.

The lab the university assigned them for the analyzer had been used as storage for years. Reese and Pearson worked hard to bring it into shape. "I'm really proud of this space," Reese said to me as we looked around at the vent hoods, work benches, and positive pressure spaces she had created. Bottles and flasks filled the shelves of glass-fronted cabinets. In one corner hung an emergency shower head for washing off chemical spills in case of an accident. "It has taken a lot of work and crying and laughing to get it to what it is," Reese said. "But it's working." She paused, the analyzer whirring in the background. Reese was about to move on from this lab, her degree with Pearson now finished and new challenges calling. She looked around the work benches and smiled, a little self-consciously. "This is good!"

A couple of days earlier, Pearson took me down the stairs of the Anderson Building, through a door leading outside to a cement pad. The pad had that back-of-a-building feeling where you expect to see heating and cooling systems and a dumpster or two. The fans of a large walk-in freezer rattled away in Auke Bay's fifty-degree air. Pearson unlocked the freezer door and propped it open with a cooler, and we went inside. She pointed to several boxes and Styrofoam packages taped securely shut with her name on them. We grabbed a couple and took them to a table outside.

"These are the samples," Pearson said as she carefully unpacked the first box. "Nobody has ever asked to see these before." I had been expecting gallon buckets of cloudy seawater. But the samples came in an assortment of plastic bags and different-colored vials. Each container had the name of a whale species and a collection date written on the outside. The fact everything was frozen cut down on what would likely have been a wicked smell.

"Somehow people find out I'm collecting this stuff," Pearson said. Biologists approached her at conference hotels to hand over samples. Colleagues brought it back on airplanes. She had several tubes sent to her by a pathologist who performed whale necropsies in Anchorage. "This is good stuff here," she said as she grabbed another package. "It came from the Alaska Whale Foundation. They collected it from a mother-calf humpback pair." The foundation, located on the east side of Baranof Island, had crews on Chatham Strait every day looking for whales. Pearson had set them up with a plankton net to sieve through the water when they came upon some waste.

It turned out this was not an easy science to practice. The humpbacks Pearson tracked around Juneau feed mostly on herring. This made it hard to collect samples because their fecal matter quickly sank. In places where the whales feed on krill, the waste has an orange color and tends to float, making it easier to scoop up from a boat. If you were lucky enough to spot a whale defecating, the collection did not need to be high tech. You could simply dunk a plastic bucket over the side if you got the opportunity. All the excess seawater is extracted by the analysis process. Even so, the chances of being in the right place at the right time were slim.

Although sampling the poop of free-swimming whales was not easy, Pearson's work on the whale pump had just received a substantial vote of confidence. The week before I arrived, she learned a German nonprofit, Whale and Dolphin Conservation, had funded the next steps of her nutrient study. She and a new graduate student would work with the Alaska Whale Foundation to put numbers on the nutrients humpback whales brought to the surface. And here is where the story gets particularly interesting: If whales turn out to be important for nutrient cycling, this would have big implications not just for the marine ecosystem but also for us. To nail down those implications, Pearson found herself getting drawn more and more into the hot new field of "blue carbon."

* * *

In July 2012, a specially created entity known as the Haida Salmon Restoration Corporation (HSRC) ran an experiment off the island of Haida Gwaii. They hired a California entrepreneur named Russ George to help them tweak the ocean chemistry off the coast. With the corporation's financial backing, George dumped 120 tons of iron sulphate and iron oxide from the back of a boat into the waters off British Columbia. The iron dump, they hoped, would be good for salmon. The sprinkling of an essential nutrient would spur the growth of phytoplankton, which, in turn, would feed the marine organisms on which young salmon depend.

The Haida tribe needed strong salmon runs for both economic and cultural reasons. But salmon restoration was only part of the story. The fact that phytoplankton suck carbon out of the water was also part the calculus. Phytoplankton take up carbon as they photosynthesize, just like forbs, grasses, and trees on terrestrial environments. The small marine organisms then enter the food chain, where they are eaten by zooplankton like protozoa, arthropods, and copepods. The zooplankton are preyed on by krill and small fish, which are themselves eaten by larger predatory fish and then walruses, sharks, and whales. The carbon remains locked within millions of different-sized bodies all swimming around in the water column.

When marine organisms die, they drift like leaves to the ocean floor. There they settle into the sediments and are dusted by marine snow. Their carbon is sequestered for millennia. If you can boost the production of phytoplankton at the surface, you can also claim to be boosting carbon sequestration at the ocean floor. George and the HSRC saw this as another benefit of their experiment. They hoped to claim carbon credits for their iron dumping and sell the credit on global markets. Entrepreneurs like George were dipping their toes into the blue carbon economy—the idea that the sea can be economically valuable in the fight against climate change.

It turned out the controversial experiment in ocean fertilization was of dubious legality and the possibility for carbon credits was slim. Although it did create a phytoplankton bloom, marine ecologists were uncertain how much of the carbon actually ended up being permanently sequestered. They also worried about how a concentrated dump of iron might affect the ocean ecosystem. The project fell apart, and George and the tribe parted ways. But the experiment caused a heated ethical debate about ocean fertilization and raised interest in the possibilities of blue carbon.

Ocean iron fertilization like the HSRC's experiment falls under the broad label of *climate engineering*. Climate engineers look at the carbon problem as an engineering challenge. The planetary system becomes a machine in need of a tune up. They propose a controversial suite of technologies that range fromr turning back the sun's radiation to rapidly sucking carbon out of the atmosphere. Fertilizing the ocean with iron to stimulate phytoplankton growth is one of them. Books on the topic have titles like *Engineering the Climate*, *Fixing the Sky*, and *Hack the Planet*.

Only a tiny portion of enthusiasts for whale recovery are motivated by climate change. But you can imagine someone excited about the whales' role in nutrient cycling making a proposal to the climate engineers. Why mess with dumping iron from fishing boats when a natural solution is right under your nose? Whales add iron to the marine system through their waste in a much more usable form than what can be dumped off the back of a boat. Climate change may be a human-caused problem. But perhaps whales can come to our aid and play a small role in helping us mitigate it. Whales, to put it provocatively, could be partners in the climate-change challenge.

Australian researcher Trish Lavery decided to look at the contribution of sperm whales to carbon sequestration by calculating how much the whales boosted the production of phytoplankton. She had some pretty solid data on the numbers of sperm whales in the Southern Ocean. She knew they did their feeding below the photic zone and their excreting near the surface. She also knew their prey, mostly squid, contained plenty of iron. This meant sperm whales were likely significant actors in the whale pump.

Lavery ran the numbers. She worked out how much iron was in the squid, how much of it was excreted in sperm whale waste, and what portion of the waste stayed at the surface. Not all of it did. ("Squid beaks rapidly sink," the paper notes.) From these numbers, Lavery was able to estimate the carbon uptake by phytoplankton and the fraction of the carbon that would eventually make it to the ocean floor. She worked out that twelve thousand sperm whales create a net carbon sink of 200,000 metric tons. Sperm whales, in other words, are like rain forests in their ability to keep carbon out of the atmosphere. The reduction in sperm whale populations by whaling meant tons of carbon going into the atmosphere each year instead of sinking toward the ocean floor.

You can do the calculations with other species and other types of nutrient transport. Blue whales don't feed as deep as sperm whales, so they don't contribute as much to the whale pump. But they do migrate a long way, which means they are big players in the whale conveyor. Roman looked at the nitrogen blue whales get from krill and estimated how much ends up in the calving grounds creating phytoplankton growth. He calculated the annual carbon sink created by blue whale poop could have been 140,000 metric tons before whaling. A recent paper calculates that whale recovery in the Southern Ocean would increase the iron cycling through the system by a factor of ten.

The carbon-capturing capacity of whales is not limited to their waste. There is also carbon stored in the whale itself. Whales are the old-growth trees of the ocean. Not only does a hundred-ton whale have a mountain of carbon in its body, but it accumulates it very efficiently. Animals who spend time in colder climates evolved bigger bodies. When you are big, a smaller portion of you is near the skin than with a less massive animal. This means you retain heat better. Better heat retention in a cold environment means you can afford a slower metabolic rate. Although your food needs are larger than those of a small animal, you need less food per unit of body mass. The food eaten by a blue whale, for example, could support fifteen hundred penguins. But the combined weight of those fifteen hundred penguins would be one-twelfth of a blue whale. Whales, in other words, give you a lot more bang for your carbon buck than penguins.

When a whale dies, the carbon has to go somewhere. A few whales wash ashore, where, if they are not hauled off to a landfill, they are scavenged by everything from flies to gulls to polar bears. The majority of them sink to the ocean depths, where they create piles of decaying muscle, blubber, and bone known as *whalefall*. This bonus habitat on the ocean floor can last for decades as the bones slowly break down. At least two hundred species feast on the whales' bodies, including some that evolved specifically for the carcasses of certain whales. When everyone is done feeding, what remains is buried under marine snow and becomes part of the deep ocean sediments.

The carbon that sinks with a forty-ton gray whale is the equivalent of two thousand years of carbon falling at the background rate of marine snow. If you multiply this by all the dead whales, it starts to add up. In 2010, Andrew Pershing, then a researcher at the University of Maine, punched in the variables. He and his collaborators looked at eight species of baleen whales and estimated the amount of carbon swimming around the world's oceans in the bodies of whales was reduced by over nine million metric tons as a result of commercial whaling. Rebuilding whale populations would sink 190,000 metric tons of carbon to the ocean floor every year as the older whales died. Pershing's numbers make it clear the carbon footprint of whaling wasn't just the fossil fuels burned to catch and process them. It was also the blue carbon that failed to accumulate. The loss of baleen whales is the equivalent of burning nearly eighty million gallons of gasoline every year. Rebuilding whale populations would claw some of this carbon back. One estimate suggests $2 million worth of climate harm would be avoided for each great whale.

Pershing is quick to point out this wouldn't solve climate warming on its own. But it wouldn't hurt. "I think the best way to think about it is that we want to conserve whales anyway," Pershing told me. "It's not going to make up for all the hundreds of gigatons that we put into the atmosphere, but that's its own problem." As we start reducing the amount of carbon from burning fossil fuels, he said, the natural carbon uptake in species like whales become a more interesting percentage of the whole. His team calculated the recovery of blue whales alone would be like preserving 43,000 hectares of temperate rainforest. And compared to ocean iron fertilization from ships, it was a no-brainer. "Our calculations suggest that rebuilding whale and fish populations would compare favorably with these schemes," he wrote. And it all happens without an engineer in sight.

Many of the scientists who work in this area admit the numbers are relatively small compared to the overall scale of the carbon problem. Pearson told me she would be horrified if the talk of ocean fertilization by whales slowed down the transition to electric cars or clean power. But all the marine biologists I spoke with liked a natural approach much more than the idea of throwing iron off the back of a boat. "Anytime we can use these nature-based solutions it's so much better," Pearson said. "Whales have been doing this for thirty-five million years." Treating whale recovery as one small piece of the response to climate change looks like a win-win-win—a win for the whales themselves, a win for carbon capture, and a win for the overall health of the oceans. Rebecca Giggs, the Australian whale expert, suggests we should switch mindsets and think of recovering whales as working on our side against climate change. Whales, she says, are "gardeners in the greenhouse."

There are signs the public is beginning to latch on to the idea of blue carbon. Pershing's paper about whales and carbon capture is now a decade old. He told me more people had reached out to him in the last year than in the previous nine years combined. "I think the policy discussion on natural carbon solutions has really advanced," he said. He is under no illusion that marine organisms will solve the climate problem. But he hopes the monetary value of the carbon they sequester might somehow find its way back into conservation.

Pearson confirmed it has taken a while for the ideas to reach prime time. "The first proposal I put out to work on the whale pump was in 2012. It finally got funded nine years later." What today's burst of interest shows is whale conservation and climate change are now firmly in the same conversation. What's in the whales' interest is also in our own.

A cynic might wonder if it takes a looming existential crisis for humans to take the interests of other animals seriously. Sometimes, no doubt, it does. But there is a different way to think about this overlap of interests, one instructive for those seeking a shift in attitudes toward wildlife. We could think of whales as partners in the climate change struggle. We don't, of course, get the chance to talk strategy with whales. Yet we are clearly on the same team. Allowing whales to multiply is good climate policy. This different framing earns whales a new type of ethical consideration, one I was beginning to see has promise for relations with other species of recovering wildlife. Because when it comes to Southeast Alaska's marine environment, it turns out whales are not the only climate partners we have.

3 HAIRY POPCORN

Willoughby Peterson greeted me promptly at 9 a.m. inside his small art gallery near St. Michael's Russian Orthodox Cathedral on Sitka's main street. Peterson and his sister were getting ready for the grand opening of their business, Xút'aa Hídi, the only Native-owned gallery in town. Their father, who is helping them with the shop, is a Tlingit carver. Tlingit means "people of the tides." Peterson pointed to one of his father's works, a beautiful six-foot-long cedar totem, on display along one of the walls. After five decades as an artist, Peterson told me, his dad still considers himself to be learning the complicated native forms.

Peterson and his sister are on a mission. Most of what the tourist shops sell in Sitka as Native crafts are knockoffs from China and India. The siblings plan to do things right. "We need to show that we know who we are and we are respectful of the art and the people," Peterson said as we looked around the gallery. "Respect is the main aspect of our culture—respect for the land, the people, the language, the spirit, nature, all animals, all things."

Peterson is slim and in his early thirties with short, closely cropped dark hair. Straight out of high school, he moved to California to take a break from small-town Sitka life. He got a degree in environmental engineering at the University of California at Irvine but did not stay long. He missed Alaska and moved back to Anchorage shortly after graduation. Anchorage, he soon realized, wasn't Sitka. "There's not that many big trees, the ocean is flatter, and it's gray," he said. "I needed more color. I missed the Tongass National Forest." It was clear pretty quickly that his move back was not just about scenery. Between leaving for California and returning home, Sitka had become a hotspot of Tlingit revival. "Cultural suppression took place for many decades, and we are just waking up out of the woods," Peterson said. "I wanted to be here to witness that."

I had sought Peterson out because the Tlingit perspective on marine mammals was essential to understand another wildlife recovery in the region. Though Peterson was quick to point out he did not speak for every member of his tribe, our conversation was full of clues about the Tlingit's distinctive way of relating to animals. Peterson started by telling me in detail the different elements of his name. He is known among the Tlingit as Goon Nax Jidi, which means "person that came from the north." Tlingits are in the Raven or Eagle moiety, a label that passes through the mother's side. Peterson is Raven. His clan is L'uknax.ádi (coho). The clans in the Raven moiety, I learned, tend to be named after animals low on the food chain (frog, dragonfly, beaver). In the Eagle moiety, they tend to be named after the big meat eaters (eagle, shark, wolf). The next subdivision below clan is house. Peterson's house is Daginaa Hít, which means "out in the ocean salmon box." Bentwood salmon boxes, he explained, were made by steaming thin planks of spruce or cedar and bending them 90 degrees to form corners. He showed me a book he recommended I read on the technique. He told me of his mother's adoption into the tribe and the ways he began his own journey toward tribal identity.

This was a lot to learn within five minutes of meeting someone. But the details of a Tlingit person's name, I quickly realized, are not superfluous. They are at the heart of how the Tlingit make sense of their place in the world. By introducing himself this way, Peterson was grounding himself in his home territory. The place I met Peterson was not the same as the one the Tlingit first encountered when this verdant swathe of coastline emerged from the ice over ten millennia ago. It had been through some dark days. But now it was in the middle of a dramatic wildlife recovery and Peterson was there to witness it. Like the recovery of whales, the recovery had significant implications for climate. It also came with its fair share of conflict. I had run into the conflict just ten minutes after arriving at the Sitka airport.

Driving into town, I found myself unexpectedly in the middle of Sitka's Fourth of July parade. I hastened off the road into the first available parking space to avoid accidentally becoming part of the procession. When I got out of the car, I saw I was parked next to a giant blue Dodge pickup truck covered in bumper stickers. One particularly prominent sticker caught my eye: "Save a Shellfish, Kill an Otter."

It was exactly the conflict I had come looking for.

* * *

Sixty years ago, you would have been lucky to spot a sea otter in Alaska. If you took a bumpy plane ride to the windswept Aleutian Islands, you might have seen a few otters floating in the kelp off a rocky promontory. If you dipped your kayak into one of Prince William Sound's glaciated bays, you would probably have glimpsed a few hairy faces, dripping water off the ends of their whiskers. But along most of the Alaska coast—between the eastern tip of Kodiak Island and the Kenai Peninsula, from Yakutat Bay down to Dixon Entrance—you would have looked in vain for *Enhydra lutris* floating on its back tucking into a clam. Like whales, bison, and beavers, sea otters were exploited to within a hair's breadth of extinction by colonial powers and the settlers they left in their wake.

Sea otters fell victim to a raging torrent of exploitation because of their coat. They have the densest fur of any animal on earth with a million hairs per square inch. This compares to seven hundred hairs per square inch on a human head. A sea otter's fur is fifteen times as dense as the beaver's infamous pelage and makes up a quarter of their bodyweight. The reason for the world-beating pelt is thermal. Sea otters—unlike seals, whales, and porpoises—lack an insulating layer of blubber. They rely on fur alone to maintain the 37 degrees Celsius they need to stay warm. Evolution delivered handsomely. The pups are born so fluffy they cannot sink. Not only must their fur be dense, but sea otters have to constantly groom their coat to keep as much air as possible between their skin and the surrounding water. A flexible bone structure lets them reach every inch of their body with their paws and their mouth. If a sea otter's fur gets matted, its insulating properties evaporate. Oil spills, like the massive one in Prince William Sound caused by the *Exxon Valdez* in 1989, are a death knell.

Ceaseless attention to their fur is not enough to keep them warm. Sea otters also need a fast metabolism to create the energy to stave off hypothermia. They eat up to 25 percent of their bodyweight each day. That's equivalent to an adult human eating over thirty pounds of ground beef and chicken between sunup and sundown. Their supercharged metabolism generates so much heat that captured sea otters have to be put on ice when taken out of the water, earning them the nickname "swimming furnaces." Even with all the food they devour, they still have to watch their energy. They rest at the surface for almost half the day to conserve calories.

A sea otter's diet varies depending on where it lives but is made up mostly of marine invertebrates plucked off the ocean floor. They eat over 150 different species, including clams, abalone, sea cucumbers, and urchins. They tear into crab, starfish, and octopus when they can find them. They also take the occasional seabird. Since sea otters are unable to dive deeper than three hundred feet, they are forced to live close to shore. They rarely cross deep channels because the depth would prevent them diving for food during the journey. This restricts sea otters to a small home range, something with consequences when too many are killed in one area.

Sea otters used to live from the Baja Peninsula up the west coast of North America in a long arc to the Russian far east and Japan. When Vitus Bering's expedition limped home to Russia from Alaska in 1742, it had nine hundred sea otter pelts on board. After people saw the quality of the fur, a commercial rush was on. Demand quickly spiked, with many of the pelts shipped to China. The Russians, who employed skilled Aleut hunters to catch the otters from kayaks, called the fur "soft gold." They hunted them from boats and with guns from purpose-built towers on shore. American vessels soon joined the slaughter and were taking ten thousand pelts a year by 1800. Their astronomical value was a tempting lure. Four sea otter hides could buy a house in Victoria, British Columbia, in the early 1900s. The commercial harvest continued until sea otters gained protection in 1911 at the signing of a fur seal treaty between Russia, the U.S., Britain, and Japan. Records for the total harvest are scarce, but fur traders likely killed close to a million sea otters during a century and a half of exploitation in Alaska. It is said the Russians sold Alaska to the U.S. in part because so few otters remained.

When the slaughter stopped, there were thirteen small populations left, adding up to barely a thousand animals. More than 99 percent of all sea otters had been killed. Most of the surviving animals were off the Aleutians and the Alaska Peninsula, with one outlying population hanging on in Central California. For the first time in more than a million years, Pacific coasts fell quiet to the rhythmic sound of otters cracking the shells of bivalves on rocks they balanced on their bellies.

* * *

It should come as no surprise that the first step in the sea otter's recovery was to stop killing them. After the 1911 treaty, the population immediately stopped falling. By the mid-twentieth century, it had rebounded to around thirty thousand. This was a good start, but it was still only a tenth of historic numbers. Without the ability to disperse far on their own due to their reluctance to cross deep channels, sea otters were unlikely to refill their original range. So in the mid-1960s, biologists decided to relocate otters from the Aleutian Islands to suitable habitat in Southeast Alaska, British Columbia, Washington, Oregon, and California. Southeast Alaska saw just over four hundred otters returned to seven sites from Yakutat to the southern tip of Prince of Wales Island.

The reintroductions were a huge success. The population grew at rates well north of 20 percent in some areas. Southeast Alaska now has over 25,000 otters. The eighty-nine otters shipped to British Columbia have grown to eight thousand. In Washington, the fifty-nine animals brought back to the Olympic Peninsula have multiplied more than thirtyfold. (Oregon, for reasons unknown, did not have the same success). The California population has increased from around fifty to more than three thousand. When these are added to the populations in the Aleutians, there are over 125,000 sea otters alive today. More than a third of them live in areas where they have been reintroduced. The success is widely regarded as a triumph in marine conservation.

There is one anomaly to the good news that has marine mammal specialists concerned. The Aleutian Archipelago was always an otter stronghold. It is so remote it provided a refuge for sea otters during the fur trade. After protection arrived in 1911, it was the first place otters began to bounce back. Recovery prospects in the Aleutians got even better when President Taft designated the Alaska Maritime National Wildlife Refuge in 1913. When biologists needed surplus sea otters in the mid-1960s for the reintroductions, they went to Amchitka Island in the Aleutians to collect them.

Having once been the epicenter of recovery, sea otters in the Aleutians are now suffering their own decline. More than eighty thousand otters—over 90 percent of the population—vanished between 1990 and 2010. The reasons are not completely clear, but the suspected culprit is killer whales. Killer whales rarely ate sea otters in the past, but they seem to have developed a taste for a snack marine biologists call "hairy popcorn." The Aleutians are so inaccessible that people don't often see killer whales actually eating sea otters. But a number of converging pieces of circumstantial evidence make them a prime suspect in the otters' decline.

The argument comes together like a maritime whodunit. If the otters were dying of starvation or disease, researchers would see skinny and sick sea otters hauled out on rocks everywhere. But they don't. Another clue is that young and healthy radio-tagged otters are disappearing suddenly and without explanation. Additional evidence is found in noticeable changes in sea otter behavior. Sea otters in the Aleutians are spending their time closer to shore, preferring to hang out in protected bays. Otters in Clam Lagoon, a two-mile enclosed ring of ocean inaccessible to orcas, are flourishing. Less than a mile away in open water they are rare. These bits of evidence are strongly suggestive of predation by killer whales. And not all the evidence is so circumstantial. A necropsy on a washed-up killer whale in Prince William Sound found five sea otters digesting in its stomach.

Experts think orcas started eating sea otters when their traditional prey collapsed. Killer whales earned their name by attacking calves of the great whales. When these whales were hunted close to extinction, the orcas were forced to eat more seals and sea lions to fill their bellies. This worked for a time until their new prey suffered its own population crash. Some marine biologists think the killer whales might be to blame here too. Others point the finger at fishermen and changing ocean conditions. Whatever the cause, killer whales had to adapt again. Hairy popcorn in the form of sea otters plugged the gap. This took a rapid toll on their numbers. Modeling shows it would not take long for a handful of killer whales to decimate the entire Aleutian sea otter population.

If this account is accurate, then the recovery of great whales and Aleutian sea otters may ultimately be tied to one another. As more whales return to the Gulf of Alaska and the Bering Sea, killer whales will have the chance to revert back to their original prey. This will relax the predation pressure on sea otters. An elder statesman of sea otter research who has studied otters in the Aleutian Islands since 1970 agrees. "If someone were to ask me what's the most likely way of fixing this problem," says U.S. National Academy of Sciences member Jim Estes, "I'd say recover the whales."

Recovery prospects for both species are improved by the fact sea otters and whales are both wildly popular animals. If whales have earned wide public affection for their charisma and reputation as gentle ocean giants, sea otters are not far behind. It is hard to resist a playful bundle of fur that holds hands with its young, uses tools to smash open shellfish, and wraps itself in kelp while sleeping so not to drift away. Sea otters are loved by almost everybody. But *almost* everybody is not enough. Although their recovery has delighted many, it has enraged a small but influential segment of the ocean community.

* * *

When I first ran into Kristy Kroeker, she was squelching along the sidewalk outside the Sitka Sound Science Center wearing a diver's dry suit. She had water droplets falling from her hair and looked cold. Kroeker and her students had just returned from a dive in the chilly waters of Sitka Sound. Kroeker runs a marine science lab at the University of California in Santa Cruz where she studies, among other things, how climate warming and ocean acidification will affect the marine ecosystem. At a research site here in Alaska, she is trying to understand how the major players in the nearshore environment might change as temperatures warm. These players include sea otters, shellfish, sea stars, and various algae and grasses. They also include scuba divers.

You don't have to be on this part of the coast long before the conflict between divers and sea otters comes up. In the mid-1960s, before the first reintroduced sea otter had wet its whiskers in Southeast Alaska, the region became home to several lucrative dive fisheries. For three decades, Alaskans donned masks and flippers and descended to the ocean floor close to the tideline to pluck abalone, geoducks, and sea cucumbers from rocks washed by Alaska's productive waters. They were a tasty treat for the family back home, and they also commanded a handsome price when sold to markets in Asia.

At first, the divers made a good living. But as more people took part, harvests declined. By the late 1980s, the dive fisheries were struggling. The Alaska Department of Fish and Game started restricting some harvests and then closed the abalone fishery completely in 1996. Pinto abalone became so scarce in Southeast Alaska they were considered for endangered species listing. The divers knew exactly who to blame. Sea otters had been reintroduced to the area in the mid-1960s. Shellfish started to decline soon after. The rats of the sea and their voracious appetites were the problem. Part of Kroeker's work is to find out if the divers were right.

A couple of days after my first encounter with Kroeker, I met with a Ph.D. student of hers who is closing in on some answers. Lauren Bell grew up in Homer, a fishing town six hundred miles northwest of Sitka. She met Kroeker during her undergraduate days at Stanford. Trading

Palo Alto's year-round sun for something considerably more bracing, Bell trekked north to Fairbanks, where she earned a master's degree in marine biology. She moved to Sitka in 2015 to help the Alaska Department of Fish and Game develop the region's first long-term abalone monitoring strategy. When Kroeker was awarded big grant on coastal ecosystems and climate change, she asked Bell to join the team.

Bell and I sat down to chat on a large rock under a tree behind the Science Center. The scrape and chime of diving gear being moved around storage sheds by Kroeker's students created the background soundtrack for our talk. As tourists peered into salmon hatchery raceways nearby, Bell told me there is little doubt sea otters took a toll on shellfish. But the relationship is much more complicated than it appears. Bell went straight for the elephant in the room. "Abalone crashed during the heyday of the commercial fishery," she said. It seems highly unlikely divers were free from blame. Although human harvest pressure is part of the picture, there is another variable that nobody realized at first was critical to the story. That variable is kelp.

Kelp is a type of seaweed with a preference for rocky coastlines bathed by cool waters. It is not a plant but large brown algae. It has fronds (or blades) the width of a kayak paddle with crinkly edges and a ferny look. A kelp blade is slick to the touch and seems to slither away from you when you try to pick it out of the water. Kelp fronds wash back and forth in the water column as waves roll over them on their way to shore. The macroalgae is prolific. With sufficient light, a suitable pH, and enough nutrients in the water, kelp can grow explosively. Some species add nearly two feet each day. The resulting forests of brown macro-algae can fill a bay in a summer.

Kelp is crucial to the health of the nearshore environment. "When you take kelp away, it reduces biodiversity," Bell told me. "The productivity just plummets." Like phytoplankton, kelp is an important primary producer. By converting sunlight into sugars, it provides food for starfish, crabs, and urchins. Mussels and barnacles growing in kelp forests are larger than those on bare rock. The waving forests of seaweed act as nursing grounds for a wide swatch of marine life.

Kelp also makes shallow waters more hospitable. A kelp forest purifies and oxygenates the water. It buffers waves before they crash into shore. The part of the kelp's anatomy that attaches to the rocks, known as the *holdfast*, traps harmful sediments running off the land. The dense forests of waving fronds create a complex habitat in which small fish can hide. The shade they cast benefits coralline algae, an important food source for urchins, mollusks, and fish. Kelp is the scaffold from which a whole mass of coastal life hangs. Charles Darwin ranked kelp above tropical forests for its contribution to biodiversity. "I have not really heard about anyone not wanting kelp forests," Kroeker said. "It powers the food webs out there." Kelp performs so many useful functions it may have changed the course of history in the Americas. According to the kelp highway hypothesis, Native people migrated down the west coast of North America through the narrow gap between the ice fields and the ocean thanks almost entirely to the productive environment created by kelp.

Sea otters, it turns out, are essential for maintaining kelp forests. Jim Estes—the same Jim Estes who blames killer whales for the otter decline in the Aleutians—discovered this in pioneering work for the U.S. Fish and Wildlife Service in the 1970s. A paper he wrote with John Palmisano on otters and kelp became a classic study for how predators exert a strong top-down influence on an ecosystem. Like wolves in Yellowstone, sea otters, through their influence on kelp, completely reshape the territory. This happens because of the sea otter's phenomenal appetite for urchins. Most of us hear the term *sea urchin* and think immediately of spines. We are less familiar with the idea urchins can devastate a healthy marine habitat. They don't look like voracious eaters, but left unchecked, urchins will mow down a kelp forest and create a wasted ecosystem known as an *urchin barrens*. Otters, with their own impressive appetites, eat enough urchins to prevent this from happening.

"It is a pretty clear relationship," Kroeker told me. "Where you see otters, you are more likely to see a kelp forest." She has witnessed this up close in her own work. "In the nineties, there were no sea otters and little to no kelp in Sitka Sound," she said. It was just bare rock, urchins, and shellfish. "In the 2000s, otters came in, and Sitka Sound turned into kelp forest." Estes and Palmisano's work in the Aleutians showed that as few as six otters per kilometer could flip an urchin barrens back into a shaded kelp forest in a season or two. The cute, hairy mustelid with the legendary appetite is the Johnny Appleseed of kelp. And in a stroke of good fortune for *Homo sapiens*, the benefits of healthy kelp forests reach far beyond its benefit to fish, algae, and bivalves.

* * *

Sea otters turn out to be on the same side of the climate fight as humpback whales. Otters don't just do a favor for creatures that live in the nearshore environment. When they help a kelp forest grow, they also do a favor for us. Kelp capture carbon as they photosynthesize, just as plants do on land. And kelp grow fast.

One of the benefits of an organism that takes carbon directly out of sea water is that it helps with ocean acidification. Bell told me her research found kelp forests providing short-term refuges from acidification during their period of maximum growth at the height of the day. The decline in acidity is only temporary, so kelp won't provide a permanent solution to ocean acidification. But they can provide a buffer. This, in itself, is good news. But the real reason for excitement around kelp when it comes to climate change is not acidification but carbon capture. Kelp grabs carbon out of sea water and embeds it in the tissue of its fronds. Kelp fronds are a much safer place to put carbon than the atmosphere.

Across campus from Kristy Kroeker's lab at UC Santa Cruz, Chris Wilmers has tried to put numbers on how much sea otters can help with the carbon problem. Wilmers is cut from the same cloth as Joe Roman and Heidi Pearson when it comes to climate change. He thinks animals need to be factored into the carbon equation, making the case in papers with eclectic titles like "Animals and the Zoogeochemistry of the Carbon Cycle." Wilmers has an eye open for how the arguments about carbon capture can support wildlife conservation.

In 2012, he and four collaborators conducted a study in sea otter habitat from Vancouver Island to the Aleutians. They wanted to quantify how much sea otters contribute to carbon uptake through their influence on kelp. They clipped and weighed kelp at sites with and without otters, calculated the difference in carbon content in each region, and multiplied the answer by all the potential kelp habitat along the coast. They found sea otters locking up between 87 and 172 grams of carbon per square meter in the ocean ecosystem. Wilmers calls this carbon bonus "the predator effect." Multiplied across all potential sea otter habitat from southern Vancouver Island to the end of the Aleutian chain, this came to between four million and eight million metric tons of carbon that could be captured by the work of sea otters. At prices on the European Carbon Exchange when Wilmers wrote the paper, hundreds of millions of dollars of carbon could be taken out of circulation by the appetites of otters. This money, Wilmers noted, would fund an impressive amount of marine mammal conservation.

This all looks good on paper. The question, Bell reminded me, is whether the fast-growing algae can sequester carbon for the long term. Like leaves in a deciduous forest, kelp die back in the fall when daylight starts to diminish. If the detritus is not locked up somewhere permanently, it doesn't do much for climate change. "Do we sink it? Do we bury it? How do we get rid of it?" Bell asked. Wilmers and his colleagues took a wild guess and estimated between 1 percent and 50 percent of the carbon in kelp would naturally end up in the deep ocean. This is equivalent to the annual emissions of up to five million cars. Not a bad climate payoff from the efforts of a cute, clam-eating mammal with thick fur and webbed feet.

Despite the gaps in the science, both Bell and Kroeker think kelp has a role to play in a warming world. It is easy to grow. It makes use of nutrients already in the water column. All you need to do is hang stringers covered in kelp spores in the water and come back in a few months. "The kelp itself," Kroeker said, "is fairly resilient and seems to bounce back from perturbations. It gives me some hope there is a lot of resilience in the system." Bell shared Kroeker's sense that kelp has a future. "There is a lot of interest in seaweeds because they potentially could be benefitting from climate change in the ocean," Bell said. Kelp profit from elevated carbon levels just like tomatoes in a greenhouse grow bigger when you pump in extra carbon dioxide, a phenomenon known as *carbon dioxide bathing*. Warming waters won't harm the kelp here either, at least not for a while. Alaska is at the northern end of kelp's natural range.

Bell and Kroeker's view was shared by an expert I spoke with at the local university. I met Angie Bowers at a bench in the same scenic parking lot where I began my time in Sitka. Bowers used to run a local fish hatchery at one end of Sitka Sound before becoming an assistant professor of fisheries at the college. "There is a lot of excitement and talk about mariculture and seaweeds in our region, but the markets aren't developed yet," Bowers said. She thought it was only a matter of time before that changed. The state of Alaska has realized commercial kelp farms could be important in the state's changing economy and is pumping money into seaweed through grants and incentives.

Bowers is confident seaweed farming is about to take off. "There are a lot of people interested in doing it," she said. "We are just getting started." Kelp has potential as a food source for people and livestock. It's uses range from kelp flour to beer. Bowers has been looking at what else kelp farming could do for the local economy. She is running experiments to determine whether kelp can be used to clean the water around the state's fish hatcheries. She has visions of salmon hatcheries surrounded by carbon-sucking kelp plantations, interspersed with crab ranches and mussel farms. "We are going to need to eat out of the ocean, and we are going to have to grow it ourselves," she said. Kelp were going to be the critical factor in making it all work.

Whether or not ocean farming takes off in Alaska, sea otters are the lynchpin of any kelp renaissance taking place in the wild. You need sea otters to keep the urchins in check. But sea otters don't eat only urchins. They also eat abalone, mussels, and crabs—hence the angry bumper sticker I saw on the blue truck when I first arrived. It looks like an impasse and a recipe for decades of fighting.

But the more Kroeker dives in Sitka Sound, the more she thinks the story about sea otters and shellfish is too simplistic. There have been signs recently of abalone and other shellfish coming back despite the presence of otters. Macroinvertebrates, such as the shellfish so prized in the region for their taste, thrive in areas high in kelp detritus. The rocky habitat on which kelp grow provides plenty of cracks and overhangs for abalone and other shellfish to hide. Another phenomenon that Kroeker has noticed is that nowhere do you find wall-to-wall sea otters. As a result of their fast metabolisms and inability to travel far, otters populate the marine landscape in patches. They can be thick in one area and completely absent across an adjacent channel. Neither Kroeker nor Bell thinks otters are free from blame for the shellfish decline. But they both think the story should be more nuanced. They are looking for a path forward beyond the bumper stickers and the vitriol.

"We are wondering," Kroeker said, "if there is a sweet spot where you can have it all." And it's talk of a sweet spot that brings the story right back to my conversation with Willoughby Peterson in his art gallery.

* * *

Northwest coast tribes have a twelve-thousand-year history with sea otters. Otters were used for food, clothing, and trade. Pictographs on cave walls in Kachemak Bay portray sea otters impaled by harpoons. Their likeness was carved into walrus ivory and used as toys and jewelry. The meat was a source of iron in tribal diets, and the fur was turned into ceremonial robes and bedding. Hunting was always done cautiously. Tribes allowed only respected chiefs and experienced hunters to participate in sea otter expeditions. They hunted from canoes with cedar arrows, harpoons, and clubs. They polished the hulls of their boats so they could slip quietly through the water without spooking their prey.

At the same time as they harvested sea otters, coastal tribes also relied heavily on shellfish. Piles of shells heaped into middens at village sites indicate the Tlingit and other Northwest tribes gathered hundreds of clams, mussels, and abalone at low tide. The mix of shells reveals the make-up of tribal diets. The middens are laced with clues. Shell size discloses whether there was heavy otter predation on shellfish. (Where otters were present, the shells are smaller.) The chemical signature within otter bones in the middens reveals whether otters were feeding on shellfish or finfish at a particular time. More finfish meant lots of kelp and lots of otters. More shellfish meant fewer otters but plenty of urchins and other invertebrates. Scientists have used these various strands of evidence to create a picture of the changing coastal ecology through time.

The evidence suggests otters were distributed across the coastal environment in patches. In some areas, they were common and the kelp forests flourished. Nearer the villages, they were mostly absent and shellfish were dominant. The pattern worked to the tribes' advantage. They had shellfish close to town, but there were plenty of otters providing their slew of ecological services within a short paddle. All of this was deliberate. A system of proprietorship gave different clan chiefs responsibility for maintaining the productivity of different sections of coastline. "Traditionally," says Tom Happynook of the Nuu-Cha-Nulth on Vancouver Island, "we had people who were responsible for taking care of the resources. And when the sea otter was abundant in our territory, this family would be charged with protecting our clam beaches." Tribes harvested more otters in areas where they wanted to boost shellfish numbers. But they made sure otters survived elsewhere. Neighboring chiefs talked with each other to prevent sea otters from being overharvested.

The fact tribal people had a deliberate management regime for a complex predator-prey system should come as no surprise at all. Tribes skillfully manipulated a range of marine resources to their advantage. They maintained herring numbers by transplanting egg-laden hemlock boughs into bays where herring were scarce. They moved abalone and salmon to restock beaches and creeks where the harvest had thinned. They created clam gardens by building rock walls between high and low tide. Similar care went into the relationship with otters. The result was a marine ecosystem that displayed what some have called a "cultivated abundance."

Researchers at Simon Fraser University and the University of Victoria have collaborated with Indigenous people to understand how the management regimes worked. They combined archaeological evidence with dozens of interviews with tribal members and learned that an effective system for otter management was in place all along the central coast of British Columbia. Harvests were tightly controlled. Rules were made, permissions sought, and transgressors were punished. Adjacent tribes worked together to support their common interests. K̲ii'iljuus Barbara Wilson of the Haida tribe told the researchers, "People were managing sea otters to protect different food sources. There would have been natural ups and downs in that time also—shifts in abundance. There would be negotiations between areas to offset resource imbalances." The bottom line for everyone was to preserve food security. "It was expected that you would be sustainable," said Guujaaw, a hereditary chief of the Haida. The otters were treated as thoughtful and self-aware participants in a shared landscape. HappyNook's grandfather told him hunters would anchor a couple of dead otters at the mouth of a bay to dissuade others from approaching shellfish beds. Deborah Head from Prince of Wales Island says an understanding emerged between people and otters. The tribes treated them as a valued part of their ancestral heritage. "You had better go to the outside waters," tribal members would say to otters swimming near their canoes, "or you are going to end up as a headman's headdress."

Through this careful management, Indigenous people maintained the sweet spot Kroeker is hoping can be recreated today. They knew how to maintain both otters and shellfish. The tribes still have that knowledge, underutilized and unappreciated—something the Canadian researchers found was a source of immense frustration. The hope is that by taking more responsibility for otter management, the tribes can help improve food security for all coastal people. There is one obstacle, however, that might stand in the way of this happening. Getting from where things stand now to the sweet spot requires an ethical pill some find extremely hard to swallow.

* * *

Forty-year-old sea otter hunter Peter Williams describes his work in cultural terms. "My process perpetuates an Indigenous protocol of developing a personal relationship with 'the materials of place,'" he wrote in *First Alaskans* magazine. "I source by hunting and fishing; I alter by tanning; I construct by the time consuming yet cathartic labor of hand sewing, each stitch a prayer." Williams hunts sea otters and seals around

Sitka from a small skiff with a rifle. When he sees an otter within range of an island, he pulls his skiff up on the far shore and creeps stealthily into position. He rests his rifle on a rock and asks the otter for the gift of its life before squeezing the trigger.

For Williams, there is no conflict between preserving the sea otter and hunting it. "Hunting a sea otter while in the realm of conservation is not seen as a conflict, but the opposite," he wrote. "It's through a constant, holistic, dependent relationship that involves both the death and appreciation of the sea otter that ensures it will be around for many future generations." Successful coexistence requires giving. But it also involves taking.

Willoughby Peterson gave me a similar message in his art gallery. "We believe these things exist for our benefit," he said. "It is only when you overhunt or overharvest that it backfires. To leave it and not utilize it with respect would also be a disrespectful thing. So it is a balance." Conservation happens through ongoing respectful use, not by walling off animals within a protected area. "It is a different perspective on how to express value," he said. "In Western societies, it seems to express value is to extract value. In our culture, to express value is to preserve it. We tend to preserve, conserve, and respect."

Coastal tribes are currently the only people in the U.S. permitted to hunt marine mammals under the terms of the 1972 Marine Mammal Protection Act. A condition on the hunt is that animals must be used for subsistence food or traditional crafts. Sea otter harvest by tribes in Sitka Sound has been growing. In some years, Native people have harvested nearly fifteen hundred otters for subsistence use.

Unpalatable as it is to many people, this harvest could be the key to creating the patchy mosaic Kroeker thinks of as the sweet spot. Otters don't range very far because of their metabolic needs. It wouldn't be hard to keep populations low in some areas to benefit shellfish and high in others to promote kelp. "The way I'm interpreting it is that you might be able to manage otters on a much more local scale than we think about," Kroeker told me. Only by reducing otters in some places can there be enough shellfish for tribal needs. The tribes not only gain traditional foods, but they also gain from the return of traditional ceremonies and practices. The patchy mosaic Kroeker envisions would mean the all-or-nothing battle lines between otters and shellfish might soften. Both could exist in shifting territories that merged and migrated as the seasons went by and the hunting grounds changed.

Such an arrangement requires a different vision of human-otter coexistence. The vision rejects the idea of people and wildlife as belonging different domains. It involves thinking of otters and people as existing in a *coupled system*. There is not one area permanently reserved for otters and one area permanently set aside for people. There is a fluid entanglement of both. Such an arrangement has worked in the past. The system evolved over millennia and created a cultivated abundance of marine biodiversity. Getting that system back is where the challenge lies. It is a challenge because the idea of a coupled system is not just a different kind of management. It's a whole different kind of ethic.

* * *

Peterson told one more story during our conversation in his art gallery that stayed with me as I thought about this coupled ethic. It was a story about his tribe's relationship to pacific herring. Herring are an important food source for people and the entire marine ecosystem in Alaska. They lay billions of eggs in early April, turning inlets in Sitka Sound aquamarine. Their young are valuable forage, eaten by numerous predators, including seabirds, salmon, halibut, and whales.

For generations, the Tlingit have cut hemlock boughs and put them in bays used by herring at spawning time. The strings of herring eggs contain an adhesive that makes them stick to the boughs. The Tlingit lift the boughs from the water soon after spawning and take the salty treat home. "A two-pound branch becomes two hundred pounds because of the eggs," Peterson said, his eyes ablaze. It's a nutritious delicacy. Tribal elders always made sure enough fish spawned to keep the harvest self-sustaining. If necessary, they transferred egg-laden boughs to empty bays to encourage new populations where herring were scarce.

The state of Alaska, according to the tribe, was endangering the traditional harvest by allowing commercial fishermen to scoop up the herring in giant seine nets before they spawned. The fishermen sold their catch to factories where workers extracted the eggs and sent them to Asian markets. The herring carcasses were thrown away or ground up for dog food. In these fisheries, known as *sac roe fisheries*, captured females never get the chance to spawn. Peterson told me the herring run had been so small in the previous two years that fishing had been completely shut down. Hemlock branches came out of the water with barely any eggs on them. Jeff Feldpausch, the director of resource protection for the tribe, said to the local radio station, "People don't want trees in their freezer. It's all about putting eggs in their freezers, not branches." The tribe filed suit.

It was while telling me this story that Peterson had made the comment about preserving value rather than extracting it. The tribe had preserved the value of the herring in the Sound through respectful use over many generations. Tribal members interacted with the herring continuously. They killed and ate some, moved others into bays that needed more fish, and always offered thanks to the herring through prayer. They knew the herring sustained many other creatures in the system. The Tlingit were grateful for the chance to live alongside them. From an ethical perspective, the value remained firmly tied to the place. At the same time, the place remains integral to who the Tlingit are, something the naming system Peterson explained to me when we first met demonstrates so well. Commercial sac roe fisheries, in contrast, extract value and take it elsewhere. Place is completely ignored.

The sort of connection between people, place, and wildlife Peterson articulated sows the seeds for a better way to think about animals when they recover. When I asked scientists about the benefits of recovering whales and sea otters for the marine environment, they spoke enthusiastically about the *ecosystem services* the animals provide. Ocean fertilization by whales and kelp forest maintenance by sea otters are clearly very useful. Ecosystem services is a language ideal for scientists and economists who are trying to capture the value of wildlife but do it without showing any passion or emotion.

I doubt the Tlingit would think in terms of ecosystem services. Such language is too one-dimensional. Peter Williams sees the relationship with otters as one of mutual benefit. "It's about connection and intimacy and having this relationship with the animal," he said to journalist Todd McLeish. "The more I'm connected with the animal, the more I work and create and celebrate with it, the more I get to know it and myself and my culture."

Peterson seemed to lean toward a similar ethic but voiced it in terms of kinship. When showing me his father's art, he said, "All these animal figures represent humans. So we see ourselves as the animals. We aren't showing the animals as animals. We are showing them as humans on the totem." In this worldview, the gap between people and wildlife dissolves. What's more, a relationship with one's animal kin must be appropriately balanced. "You can see it in our art," he told me. "The totem pole is balanced symmetrically—up and down, left and right. Our culture is balanced. So I think that's a manifestation of what we do in nature."

I thought back to Heidi Pearson's ongoing explorations of the social behaviors of whales and primates. She makes some of the same connections as Peterson but from an entirely different cultural framework. Whales, dolphins, and mountain gorillas are not alien forms of life. They are kin with lessons to teach. It should come as no surprise the conditions that benefit whales and otters—well-fertilized oceans, healthy kelp forests, abundant shellfish—also benefit us.

I think the term *partnership* also works well. We obviously don't negotiate with whales or otters as we do with our human partners, but they remain essential to our well-being. We might think of them as the kind of beings we want to have around in difficult times. If we watch and listen, we can gain copiously from their presence. Thanks to behaviors stitched into their tissues over evolutionary time, whales and otters act in ways that keep natural systems healthy. They show us the way, simply by being themselves.

Close observers of wildlife repeatedly notice this. Henry Beston spent more than a year living in a simple cabin on Cape Cod watching the ocean's shifting moods. It was therapy for the trauma he witnessed as an ambulance driver in France during World War I. In the book he wrote about the experience, *The Outermost House*, Beston concluded that nature and the wildlife it contains are part of our humanity. Without them, we lose fragments of ourselves. We need them to be whole.

Beston also said this about animals: "... they move finished and complete, gifted with extensions of the senses we have lost or never attained, living by voices we shall never hear.... They are other nations, caught with ourselves in the net of life and time, fellow prisoners of the splendor and travail of the earth."

Whales and sea otters exist in the same bewildering net of life and time as we do, living a splendor we can only glimpse during our short excursions beneath the waves and their short visits above. As their populations recover, they will join us firmly in the travail that is climate change. We would do well to appreciate our need for their company. There is a new way to think about their importance now. We need them not to turn their bodies into oil and fur coats to serve our insatiable desires. Nor are they merely something good to look at from the deck of a tour boat. We might think of them now as integral to our shared future, valuable partners in keeping the ravages of carbon at bay.

CREATIVITY AND COURAGE

I think of it as a source of good fortune that I live in a place chock-full of black bears.

Every few days this spring, an email with the subject line "UM Alerts" popped into my inbox from the university's Office of Public Safety. Another bear had been spotted on campus. Some bears were padding along hiking trails behind the football stadium. Others were spotted poking around the dining facilities in search of dumpsters or walking between classroom buildings in the early morning light. Usually, a campus police officer kept an eye on the bear until it wandered off. The email always contained the line "the bear was not acting aggressively," but it warned students and staff to keep their eyes open to avoid any surprises.

Bears in western Montana remain plentiful. Similar stories of abundant wildlife are not hard to find. Bobcats in New England have increased tenfold since the 1970s thanks to the regrowth of forests. Bald eagles in the U.S. have climbed from just over four hundred nesting pairs in the 1960s to more than seventy thousand today after a ban on DDT. Common cranes are nesting again in Ireland after three hundred years, courtesy of the rewetting of peat bogs. Saiga antelope in Kazakhstan have nearly tripled since a 2015 die-off. River otters are back in the San Francisco Bay. There are ten times more wolverines in Finland than three decades ago. Harbor porpoises are increasing 10 percent a year off the Washington coast. Northern elephant seals have recovered from less than a hundred to nearly 200,000. Europe's golden jackal population has increased more than thirty times, with jackals now breeding in France and Denmark where they have never lived before. Not everything is doing so well, of course, but the list of recoveries offers hope.

Some species have recovered on their own after a shift in how people use the landscape. Others require painstaking restoration efforts. Whooping cranes in North America have climbed from sixteen to more than eight hundred after decades of careful captive breeding. Their handlers disguised themselves as cranes so the hatchlings would not imprint on humans. They spent years trying to teach the inexperienced birds to follow ultralights from Wisconsin to wintering grounds in Florida. There were less than a hundred Iberian lynx in Spain and Portugal at the start of the twenty-first century. Today there are more than a thousand after a two-decade, transboundary effort centered around five purpose-built breeding centers. Giant otters flown in from zoos in Hungary and Sweden have produced pups in Iberá Park in northeastern Argentina, correcting a forty-year absence. Cheetahs are being restored to India from South Africa. Beavers are being reintroduced to Central London.

Stories of recovery like these create optimism. They hint at possibilities. On the surface, the possibilities are ecological and biological. More whales. More cranes. More beavers and lynx. More trophic cascades to nourish complexity and promote diversity. If the right practices are implemented and the right laws enforced, landscapes might heave with wildlife again.

Look a little deeper, and you see recoveries that pry open new ideas and attitudes. Wildlife resurgence demands different approaches to the creatures that will live among us. The contours of these approaches are becoming clear. The fanatical separation between the wild and civilized that drove colonial powers—and then their environmentalism—must soften. The obsession with a species' genetic purity common among many conservationists may, in some contexts, have to be relaxed. Overconfidence about our expertise at manipulating complex systems like rivers must be restrained. Acts of assistance may be necessary for some vulnerable species on a long-term basis. And new ethical attitudes toward wildlife, including partnership and reciprocity, must be nourished.

Many of these ideas and attitudes already exist in the cultures of Indigenous peoples. They, after all, lived thousands of years alongside a profusion of tenacious beasts, absorbing an ancient intelligence about how to share a landscape. Certainly, all is not lost. Potawatomi plant ecologist Robin Kimmerer points out that natural systems have "no goal other than the proliferation of life." If wildlife are freed from the conditions that have suppressed them, if cultures that slaughtered animals have the humility to learn from those who kept them alive, if soils and woodlands are restored to ecological health and laws taking care of wildlife are enforced, then the century ahead could look dramatically different from the last. Wildlife will know what to do.

There is, however, a shadow that tempers any optimism about animal recovery. I encountered this shadow in almost every wildlife story I followed. I would listen to an expert relay some good news about a species, and then there would be a pause as we both found ourselves drifting toward a terrifying whirlpool we had not yet talked about. That whirlpool was climate change.

* * *

The most recent assessment report on climate by the Intergovernmental Panel on Climate Change lays out some sobering facts. "Widespread and rapid changes in the atmosphere, ocean, cryosphere and biosphere have occurred," the report begins. The greenhouse gas concentrations responsible for the changes are "unequivocally caused by human activities." The consequences of these changes are severe.

Temperatures are increasing, and precipitation is becoming more violent in many parts of the world. Glaciers and sea ice are in rapid retreat. Oceans have acidified, and oxygen levels at the surface are decreasing. Atmospheric winds and storm tracks are shifting. Extreme weather events like cyclones, hurricanes, and wildfires are becoming more common. Coastlines are being redrawn as sea levels climb. The report leaves no doubt humans have become a planetary force. The amount of carbon dioxide, methane, and nitrous oxides our species has pumped into the atmosphere in the last 150 years would have flipped the whole earth from ice-covered to green in a previous epoch.

All of this is set to get worse. Feedback loops will create punishing accelerations in warming. Less snow will lead to more heat absorption at high latitudes. Melting permafrost will create increased methane emissions. The suffering will be immense. Drought, wildfire, floods, and disease will kill hundreds of thousands of people, many of them ill-equipped to face such dangers. United Nations Secretary-General António Guterres said the report was a "code red for humanity." It is. And code red for humanity is also code red for wildlife.

Countless species are already struggling to cope with the rising temperatures. Warming episodes such as The Blob have disrupted marine food webs, causing whales and sea lions to starve. The algae that bring coral reefs to life are being killed by hotter seas. Wolverines and polar bears have less habitat to hunt as snow and pack ice diminish. Insects are emerging from their larvae out of sync with the foods they require. Cold-water species like salmon and bull trout are having their energy sapped during nights in which temperatures never fall. Grizzly bears have lost vital whitebark pine seeds as heat pushes ancient forests off the tops of mountains. Ocean acidification is making it harder for marine invertebrates to form shells, something sure to have cascading implications for all marine life. A system that has spent millennia fine-tuning itself for resilience is being knocked off-kilter, its foundations crumbling.

The living world is doing what it can to survive. Biologists have tracked numerous species trying desperately to cope with the changes. Birds are laying their eggs earlier. Plants are adjusting the timing of their bloom. Everything from butterflies to fir trees have moved north, and higher in elevation, in a search of more hospitable temperatures. Some animals are shape-shifting. The bills of Australian parrots have grown nearly 10 percent since the 1870s to dissipate more heat. Masked shrews in Alaska have developed longer tails to help them with cooling. Bats in China have lengthened their wings so more of their blood vessels are exposed to the air.

These strategies are ingenious, but they all have limits. The more specialized an animal, the fewer options it has. Only a handful of species are likely to win the race against warming into which humanity has pitched them.

* * *

At the age of ninety-four, veteran BBC broadcaster Sir David Attenborough wrote what he called his "witness statement" to the decline of life on earth. The slaughter of wildlife is "our gravest mistake," he claims in *A Life on Our Planet*, a mistake the climate crisis is only making worse. Attenborough is convinced we still have the opportunity to put things right. But putting it right involves acknowledging a deep truth. The climate crisis and the biodiversity crisis are linked. "It is no accident that the planet's stability has wavered just as its biodiversity has declined.... To restore stability to our planet," he wrote, "we must restore its biodiversity. It is the only way out of this crisis that we ourselves have created."

Eliminating the use of fossil fuels is the absolute prerequisite for addressing the climate crisis. The restoration and rewilding of degraded landscapes is a crucial accompanying step. Nature-based responses to climate change are cheap and effective. And if restoring forests, mangrove swamps, peat bogs, and soils is good for carbon, it is a stunning gift to wildlife, a gift that also gives back. The role of animals in the carbon cycle is becoming increasingly clear. It is not just what whales do for phytoplankton and beavers for fire suppression. Recovering wildebeest have turned the Serengeti back from a carbon source into a carbon sink. Sharks and wolves control dugongs and caribou, helping seagrass beds and boreal forests to absorb more carbon. Toucans disperse the seeds that ensure the next generation of giant tropical trees can sprout. Simply boosting the size and number of animals leads to more carbon locked up in their bodies rather than floating free in the air or drifting as bicarbonate in the ocean. Wild animals are allies in the struggle against building heat.

We need to relearn how to live with wildlife if this partnership is to flourish. This is a practical sort of learning, but also a philosophical one.

It requires a wholesale reorientation in how to think about wildlife. We need greater tolerance for the company of nonhumans, thinking of them not as adversaries but as kin with common goals. One aspect of the mental transition involved is certain. It doesn't stand a chance of happening without courage and creativity.

* * *

I sat on Bryce Andrews's porch in the copper light of a September evening looking east toward the Rattlesnake Mountains. Cattle grazed in the field a quarter mile in front of us. Behind them a strip of pines caulked a crease at the base of the mountains. The haze from the summer's wildfires had finally begun to clear, and the outlines of each tree were crisp. An irrigation ditch gurgled on a shrubby fenceline nearby. Andrews's dog, Wren, fussed around, claws tapping on the wooden deck. Wren was excited to see his owner back from the raspberry field where Andrews had spent most of the afternoon trellising new canes.

I reached out to Andrews to hear his thoughts about the prospects for a new attitude toward wildlife. Andrews brings a seasoned perspective to the question. He works for People and Carnivores, an organization devoted to coexistence between farmers and large predators. He and his wife have a cattle ranch and a small berry farm on 150 acres on the Flathead Indian Reservation not far from where Roy Bigcrane lives. Andrews is also a writer whose work casts a critical eye on western attitudes to wildlife. His award-winning book, *Down from the Mountain: The Life and Death of a Grizzly Bear*, recounts a season toiling alongside farmers in the Mission Valley to reduce conflict with the growing population of grizzly bears. The bears had learned to enter the farmers' fields, sit on their haunches, and eat corn between their paws like kids at the fair.

I asked Andrews what has to change for things to work out better with wildlife this time around.

"Your question is a wall that I beat my head against regularly," Andrews said. Coexistence, he suggested, requires a deep questioning of attitudes in carnivore country. "We think a lot about how the animals affect us, but we think very little about how our use of the landscape affects them," he said. "We need to fundamentally reexamine what we think is a right or a wrong way to exist on this landscape."

Andrews told me his biggest frustration is witnessing ranchers unwilling to think creatively. Their talk of being bound by tradition, he says, is misplaced. Earlier generations were finding creative solutions all the time. "You come from a line of people who were experimental," Andrews wants to tell them. "Every single generation has been wildly innovative. Most traditional ranchers are not really thinking like that today." Making a living where carnivores are returning is not easy. ("Bloodless coexistence is a myth," Andrews said.) But the world in which settlers wiped animals off the landscape without shame is not the world we live in today. The calculus had shifted. New ways of thinking demand innovation, creativity, and courage.

I had heard this plea for creativity and courage expressed before. Lauren Bell's work on kelp and climate change in Alaska had convinced her that fishing communities like Sitka's have to change. "I see our nostalgia over what species we want to see persist here as negative," she told me. "It is hard to break from the investments you have made. But if you can't adapt to what's changing, it is going to be difficult." Bell admitted it pained her to think of how Sitka Sound might transform. She grew up with a picture-perfect marine ecosystem at her back door. But change is coming, and she sees opportunities for Sitka's fishing community to adapt and flourish as some animals recover and new ones arrive.

Like Andrews with the ranchers, Bell wants the fishermen to open their minds. Nothing in ecology stands still—not the animals and not the environments in which they live. "The ocean is going to adapt," Bell said, "and the question is 'How will we respond to that?' To me it feels more optimistic to be honest."

Responding to change, a Native friend reminded me, is what Indigenous people have done all along. Wildlife kept them company throughout these transformations. In the changes we face today, some species will startle us with their adaptability. We should create as much space for these adaptations as possible. Others will help buffer the system with the productivity they bring. And some will go extinct. The rest of this century will reveal which animal belongs in which category. But I'll wager a bet that Attenborough will be proven right. The climate future will be brighter if we enter it with robust populations of wild animals weaving protective ecological webs through the systems around us.

* * *

When I first moved to Missoula, I heard it was once possible to see giant chinook salmon in the mountains that loom above the Lochsa drainage. The idea one might walk among towering cedars and see a salmon in the dappled light of a mountain stream seemed too ancient to be real. The Bitterroot mountains are hundreds of miles from the Pacific Ocean. I assumed the salmon had died in the giant maw of postwar development that reshaped the northwest. Most of the talk of salmon these days concerned the obstacle four Snake River dams posed to fish that made it through the gauntlet of the Columbia. I doubted any fish still made it to the mountains. But I had never heard definitively they were gone.

Just over a year ago, I called the Idaho Fish and Game office in Lewiston whose region encompasses the Lochsa River. I asked the officer who picked up the phone if he knew the nearest place I could see a salmon.

"As a matter of fact, I do," he said. "Somebody asked me the same question recently, and I tracked down the answer." It turned out there is a small fish trap on the Lochsa River not far over the pass that marks the Montana-Idaho border. Every year between July and September, a few chinook salmon return to the facility to spawn. The fish swim into a cement channel where a biologist on remote assignment counts and measures the salmon before putting them in a tank to ripen for spawning. When the time comes, a small team assembles to take sperm and eggs from their bulging bodies. They mix the two together and transport the fertilized eggs to a larger facility a hundred miles downriver. The fish spend more than a year there before being brought back up to the trap to imprint on the waters of the mountain stream. The salmon smolts are then released into the river to begin their long journey back to the Pacific.

I visited the fish trap twice that summer and chatted to the trap tender. Sam Roetering was from the Midwest and was spending the summer working for Idaho Fish and Game. On a hot July afternoon, I found her sitting on the ground digging weeds from the gravel in front of the trap. The fish had been slow to arrive that year, she told me, but were starting to trickle in. They tended to arrive at night. Roetering hoped there would be enough to spawn by the time September rolled around. The salmon run was barely hanging on, dwindling over the years to around three hundred.

I asked Roetering if there were any fish spawning outside the hatchery this far up the drainage. I felt an urge to know if there were salmon living wild in the Bitterroots. Roetering did not know for sure, but told me she had chatted with a man named John who visited the trap frequently. He had assured her there were. I asked Roetering if she could get John's contact details if he came back.

A couple of weeks later, Roetering sent me an email that included John's phone number. I called him up, and we talked for a while. John was retired and living in Missoula. He confessed he had an obsession with the Lochsa. He made the hour-long drive over the pass more than a hundred times a year. Steelhead were his thing, but he knew where I could see some salmon. It was a few miles above where Roetering worked, beyond where most of the hatchery fish ended their odyssey. He suggested I get up there quick. The salmon were close to spawning. After that, they would die, and their carcasses would wash downriver and be pulled into the forest by scavengers.

Shortly after our conversation, I headed over the pass with John's detailed instructions in hand. I'm not going to tell you exactly where John told me to go, but it was easy enough to find. I'm also not going to tell you exactly when I went there, though you could probably work it out. At the appointed spot, a pale slab of rock created a mini-cascade that fed a thirty-foot reach of deeper water. I hadn't been standing there long before the dark dorsal fin of a big male chinook salmon broke the surface. It took several minutes for my eyes to adjust to the light on the creek, but when they did, I saw another male and several smaller females flicking their tails around the pool. On the far side, beneath four or five inches of crystalline water, sat a gravel bed covered in pea-sized stones. It was perfect habitat for a female to dig out a small depression for her eggs. The males chased the other fish around while two mature-looking females scouted the gravel bottom. Their bodies flexed in the late-afternoon sunlight to keep them stationary in the current.

I had seen tens of thousands of salmon before in Alaska, but none were this far from the ocean or this high in the mountains. It was hard to fathom how steep the odds were that these fish had overcome. They had swum more than six hundred miles and climbed over 3,500 feet since they left the ocean for the long, freshwater pilgrimage back to their birthplace. They had negotiated fish ladders on eight giant dams on the Columbia and the Snake rivers, dodging fishermen and aerial predators along the way. They had eaten very little for months as their bodies went through a final transition, their skin darkening, the jaws of the males forming the menacing hook biologists call a *kype*. Every rational bone in my body told me it was impossible they should be here. But here they were, swimming outside cement channels, returning to the mountains to act out the last few scenes of their life.

The chinook salmon I gazed upon that day contained a message about all returning wildlife. They were survivors. They had the power in their muscles and the wisdom in their genes to complete an improbable journey. Their eggs would nourish dippers waltzing along the bottom of cold mountain streams. The bodies of their young would strengthen killer whales foaming the distant Pacific with their flukes. Each fish enacted its part in a story millions of years in the making. There is a reason salmon are beacons for native people and icons for those with a love of the wild. They exist timelessly alongside us, stitching together a world in which people of conscience desire to live.

* * *

There was one more thought I had as I left the salmon behind that evening. There will come a day when earth's human population will peak. It may be shortly after the middle of this century. It may be later. Up to that moment, earth's biodiversity is grinding its way up a hill like an ancient rollercoaster, slowing so much it looks like it may never make it to the top. Each loud clack is a doubt about what will be left behind. But when it does reach the crest, the long struggle against gravity will finally flip to become an accelerating freefall. Millions of bodies will slither, fly, and hop their way back onto soils and waters released from service to human needs. Seeds will float in and take hold, rhizomes will reach outward from their strongholds, and life-giving spores will alight on surfaces previously covered by crops and cement. There will be a joyous free-for-all as uncountable numbers of fins, hooves, paws, and carapaces rush onto newly liberated spaces. As earth has done repeatedly in the past, life will bounce back.

Recovering wildlife are the reservoirs that will fill out the green and blue spaces recoloring the map. The work we do now will be a bridge to a world we can only dream of. Trickles of possibility will become cascades of life as animals regain the room to flourish. Just as we did for thousands of years in the past, our species will start to remember an ancient wisdom born of entanglement with the lives of animals. We will find ourselves bending once again to the pull of biological forces that never fully went away.

ACKNOWLEDGMENTS

Writing about wildlife and weaving a philosophical thread through the story turned out to be much harder than I expected. I needed a teeming rainforest of help.

The forest I encountered turned out to be full of generous, big-hearted people who helped me stitch together the many fragments that became this book. They answered my plaintive phone calls, said yes to my desire to meet them, showed me their study sites, and responded to numerous follow-up questions after I returned home. This book was as much a human adventure for me as a literary one. Hundreds of pieces had to be assembled to make it whole. I owe thanks for every single one.

In the Netherlands, I was helped by Hugh Jansman, Martin Drenthen, Wouter Helmer, and Ronald Goderie. Hugh's raw enthusiasm, Martin's intellect, Wouter's talent for getting things done, and Ronald's wit were all inspiring. Martin's friendship over many years is something I prize. Nathalie Soethe showed me what it means to put boots on the ground for wolves and was very generous with her follow-up time. Jan Opsomer is an old friend from Belgium with a specialty in ancient philosophy. He shared his take on Belgian wolves and on the cultural curiosities of the region's people. It was fun to reconnect, and the conversation helped me see my way to the end of the farmland section.

In Italy, Mario Cipollone and Angela Tavone were kind to me and expressed unfailing warmth, both in person and in the video chats COVID imposed on us. Alessandro Rosetti shared a portion of what he knew about chamois in Monti Sibillini National Park and was very good company along the way. Francesco Romito answered my questions about wolves in northern Italy. David Voros and Eros Tassi made the time in Umbria possible and fun.

Martin Biuw, Fern Wickson, and Svein Anders Lie turned my visits to Tromsø into fertile ground for thinking about wildlife. Trine Antonsen, Erik Lundestad, and Anne Myhr have also helped enormously. The visit to the Lofoten Islands with Trine and Erik was a big moment. The community of friends I have built in northern Norway over the years makes it feel like a second home. Fern was a rock during COVID and consistently inspired me when I was in search of big ideas.

Chasing down the barred owl story in Washington, Dave Wiens responded rapidly to my request and helped figure out the logistics of a visit. Later he submitted himself to a grueling phone interview and answered my difficult ethical questions honestly and openly. In the woods around Cle Elum, Melissa Hunt was a remarkable guide. Melissa worked around my schedule and helped me lay eyes on owls when I needed to. She was also generous with her time after the trip and carefully read a draft for errors. Her feedback was wonderful, her commitment to her work remarkable. Matt Larson, a local owl expert, answered a bunch of follow-up questions when I needed to revisit the topic. Denver Holt of the Owl Research Institute was similarly helpful. Kate Davis clued me into the world of raptors on her gorgeous land in the Bitterroot Valley. Anthony Keith and Iola Price clued me in some more. Tim Fox, whom I listened to in the Willamette National Forest, was the initial inspiration for writing about the northern spotted owl. His owl calls were spot on.

First, she put up with more than an hour of my questions on the phone. Then Alexa Whipple welcomed me to the Methow Valley and took me around some fascinating sites transformed by beavers. Alexa lives and breathes her work and does it with the enthusiasm the beavers deserve. Chris Johnson and Chris Pasley both spent time with me in the valley to talk fish, environmental values, and habitat restoration. The

Methow is lucky to have them all. Ben Goldfarb plugged a few holes in my knowledge of castorids.

Visiting the sites where the Elwha dams were removed was the fulfillment of a long-held dream. Robert Elofson from the Lower Elwha Klallam tribe was patient with my questions, generous with his knowledge, and extremely good company. Kim Sager-Fradkin and Mike McHenry answered my questions promptly and politely and made sure I laid eyes on some returning chinook salmon. Seeing the tenacity of the fish making their way through the emerald waters confirmed everything I suspected about wildlife recovery. Navarra Carr and Lindsey Schromen-Wawrin from the Port Angeles City Council filled me in on the relation between the restored Elwha and town politics. Kevin Colborn and Tom O'Keefe helped color in the bigger picture of dam removals. Rebecca Brown answered my questions about vegetation recovery at dam sites. George Pess gave me some key statistics on the Elwha's salmon. Andrew Whiteley graciously shared some of his expertise in fish genetics. Wayne and Kathy Hadley gave me a good set of reader's eyes.

Mike Kustudia and Rob Roberts were essential in helping me see up close two dam removals closer to home. Rob was juggling the complexities of the Rattlesnake Dam project at the very moment I demanded his time. I appreciate his patience and instructive commentary. Mike was generous with his insights at the Milltown Dam. He entertained me in his office and hopped on several follow-up emails and calls. The work he has done at the former dam site to create Milltown State Park will reverberate for generations. Amanda Bielby showed great openness to talk with me about art, health, and restoration, and I'd like to give a shout out to the nonprofit organization Open Air, which coordinated the artist-in-residence program Bielby enjoyed.

Brad Meiklejohn was very helpful on the phone from Anchorage explaining the situation on the Eklutna River. Andrew Lahr let me watch him gather data at his beaver dam analogue sites in the Lolo National Forest. Sam Roetering made my visits to the Powell Fish Trap in Idaho in search of salmon a rewarding quest. Kudos to the Idaho Fish and Game officer who appeared out of nowhere when I was on the riverbank to check I was not poaching salmon. Joel Patterson and Caleb Armstrong at the Dworshark Fish Hatchery answered important questions about the role played by the hatchery in salmonid restoration in the Clearwater and Lochsa drainages.

Diane Boyd brought her expertise on wolves in North America to bear at just the right point in the story. She carefully read a draft and improved my knowledge of wolf biology. Betty Thisted let me sit on her porch at her beautiful property in the Ninemile Valley as she reminisced about her late husband, Ralph. She and her granddaughter showed me the rendezvous site where the first pack to return to the Ninemile romped and ate roadkill. Rick McIntyre has given me and my students endless time and inspiration in and around Yellowstone. We, and the wolves, thank you.

Roy Bigcrane is a legend. I was extremely fortunate Roy was willing to spend time with me to talk about bison recovery on Confederated Salish and Kootenai Tribal land. Roy also shared the painful history of the Bison Range. Roy's generosity of spirit is something the culture I identify with has certainly not earned. The recent return of the Bison Range to tribal management is the very least he deserves. Amy Coffman, refuge manager at what was then the National Bison Range, gave me her time, as did Annie Sorrel, who was full of insights about biocultural restoration. Tom MacDonald of the Confederated Salish and Kootenai Natural Resources Division gave me a valuable perspective on wildlife from the Flathead. Terry Dahl gave me the same from Blackfeet country. Thanks to MJ Derosier for getting me connected. Lou Bruno, an icon of wilderness protection in Montana, also spent time explaining some history and offering some visions for the future of the Rocky Mountain Front.

Beth Saboe offered me an entry point into the world of American Prairie. Interviews and phone calls with Beth yielded a meeting with the group's vice president, Pete Geddes, and several conversations with restoration manager Daniel Kinka. American Prairie's work is pioneering and delicate. I appreciate very much their openness with me. Ryan Nemmers somehow found time in his schedule at the Heart Bar Heart ranch to spend a morning with me. He was informative and helpful. Andrew Jakes gave me some important insights into the world of pronghorn antelope and carefully read a draft of what I wrote about them.

Nancy Behnken—friend, fisherman, artist, and carpenter—helped me organize a very productive time in Sitka. Lauren Wild and Angie Bowers at the University of Alaska Southeast's Sitka campus opened my eyes to sperm whales and mariculture. My thanks to Laura Busch of the Sitka Sound Science Center, who appeared at just the right moment to help me schedule an intriguing set of interviews. Davey Lubin also helped set the Sitka scene. Will Peterson gave me an essential Tlingit perspective on the marine environment. Heather Bauscher from the Sitka Conservation Society filled me in on the local environmental battles. Ron Heintz offered an interesting overview of several research areas in Sitka Sound. Taylor White shared her knowledge of sea otters and the tribal context, making me think hard about the idea of coupled systems. Lauren Bell is as good as it gets on kelp and abalone. Our conversation helped me get a handle on the complicated science of the nearshore. Kristy Kroeker found time to squeeze in an interview while juggling several graduate students and a concentrated couple of weeks of fieldwork. It was much appreciated, Kristy. Linda Behnken talked with me about the black cod recovery. Robin Welling was kind enough to let me housesit and use her wheels. (I hope I took good enough care of the rabbit.) Eric and Brita Speck, together with Pat and Chris Glaab, made me grateful, once again, to have connections in such a magical place.

Heidi Pearson in Juneau generously poured time into an unknown visitor's research project. She helped me set the scene on Zoom before I made the trip to Alaska's panhandle and gave me all the hours I asked for when I arrived. The day on the water with her husband, Chris, and daughter, Maia, was just what I needed to populate the picture of humpback whales I was building. Her careful reading of draft chapters sharpened the message and helped eliminate errors. Heidi's work embodies all of what this book is about. I'm extremely grateful to her. Melissa-Rhodes Rees gave up an afternoon of her life to show me the lab and describe what it took to get an unusual research project off the ground. The walk up the Mendenhall River was the perfect end to our day.

Andrew Pershing answered all my questions about carbon sequestration by whales. Anne Salomon reached out with a giant helping hand in my quest to understand how to live alongside sea otters. Thank you to

Deborah Head for explaining the context of her story about her Haida ancestors and for giving me permission to use her voice in the text. Thanks to Sonia Ibarra for making the connection to Deborah. A major hat tip is due to the staff at the visitor center at the Mendenhall Glacier, the curators at Juneau's Alaska State Museum, and the staff at Sheldon-Jackson Museum in Sitka for the way their work deepens the experience of the visitor. Thanks also to Juneau for its affordable public bus system. And lifelong gratitude to the memory of Toby Rilling. You made Alaska come alive for many of us, Toby.

The interviews in the UK gave me a glimpse of the fire now blazing within one of the oldest environmental movements in Europe. Stan Smith was faultlessly polite, entertaining, and liberal with his knowledge as he walked with my sister and me around Blean Woods. The work being done by the Kent Wildlife Trust will create something of lasting significance in the region and beyond. Laura Gardner, Vicki Breakell, and Adam Roberts shared a ton of insights around a picnic bench at Wildwood and even more in a follow-up call. Jake Fiennes gave me far more than our scheduled time in an interview at the Holkham Estate. Josh Bean put up with my endless questions as we chugged out to see the seals at Blakeney Point. Sarah Ward of the Sussex Wildlife Trust helped get me up to speed on efforts to restore kelp in southern England.

My sister, Fiona, accompanied me throughout the interviews in England and let me relax in the passenger seat of her car as she battled with navigation and traffic. It was my first visit home since COVID, and the time we shared was all the sweeter for it. Seeing Norfolk for the first time with you was more than a treat. Where did I get the luck of having you for a sister? I cherish it always.

Toward the end of the writing process, I was fortunate enough to spend a month among the environmental philosophers at the University of Turku in Finland. Helena Siipi was a delightful host with a keen eye for how to improve the manuscript. I'm enormously grateful I had the chance to interact with her. Markku Oksanen was also a thoughtful interlocutor who put several hours of his time into my project. Elisa Aaltola helped steer my thoughts in the right direction. A number of students at the university—especially Laura Puumala, Jenna Aarnio, Mikko Puumala, and Oskari Sivula—added light and insight to the work. Juha Hiedanpää at the Natural Resources Institute Finland talked me through a number of local wildlife issues and let me present an outline of the book to eighty of his colleagues. Juha shared food and a memorable hike at a nearby national park, adding his wit and engaging personality to a remarkably rewarding month.

Dozens of students at the University of Montana have brightened my life and expanded my brain over many years. I recognize a particular debt to Ian Weckler, Anne Belldina, and André Kushnir for pushing me in productive ways when this book was first gestating. Eric Higgs in Victoria and Daniel Spencer in Missoula talked to me about the philosophical contours of ecological restoration. Deborah Slicer has been a wonderful sounding board for all things wolfy as well as a good friend over many years. Thank you all.

Natalie Elliot arrived in town and built our friendship at the perfect time. Natalie is a talented writer, witty conversationalist, and polymath. She showed me that writing is something certain people are *compelled to do*. She had no time for the shadows of self-doubt that would flit through the corners of my mind. She boosted my writing confidence both when I needed it and when I hadn't yet realized I needed it. My gratitude runs deep.

In addition to those already mentioned, several friends, acquaintances, and colleagues read drafts of different chapters and helped whittle the story into something presentable. I wish to thank Marta Mengs, Angela Hotaling, Barbara Theroux, Mason Voehl, Andy Rahn, Anne Yoncha, Diane Boyd, Fern Wickson, Wayne Hadley, Kerry Foresman, and Sterling Miller for the consideration they gave to my words and sentences.

Writers Ben Goldfarb, Emma Marris, Bryce Andrews, and Peter Stark have all been generous with their counsel and offered valuable clues about the craft.

Kevin O'Connor is my literary agent and deserves sincere thanks once more for his belief and high competence. Beth Clevenger at the MIT Press showed the kindness to work with me again and has been a most supportive and insightful editor.

The Atlantic printed an earlier version of "Pruning Apples for Bears" under the title "Why Italians Are Growing Apples for Wild Bears." Their editors were incisive and helpful. Humanities Montana, the Humanities Institute at the University of Montana, and the Kone Foundation in Finland all gave me financial support at key moments. This support permitted travel for interviews and uninterrupted research time in helpful settings. I appreciate their generosity very much.

I count myself extremely lucky to have parents and siblings who are relentlessly enthusiastic about my efforts. The contortions I went through to put this book together often made them smile. I hope they will smile again when they see the final product.

If there are errors in the manuscript or interpretations of the data that seem dodgy, please blame me and not my sources. People gave me their time and expertise freely and earnestly. I am sure I missed some key comments or drew some conclusions too hastily. Apologies in advance to my sources and my readers if this happened.

I suspect I'm not the only writer who feels they spend much of their time on a cliff edge strewn with emotional boulders ready to topple them into the abyss. The lucky ones among us have partners who steady them when they wobble and catch them when they start to fall. This person puts up with their pride, their defensiveness, and their vulnerability. They also get to join the celebrations and share the occasional joys. My wife, Lisa, not only rode the rollercoaster of this book with me, but she read every word of the manuscript, many of them multiple times. She found the cluttered sentences, pointed out my biases, and put up with my not-always-admirable reactions. She was a fountain of good suggestions about readers, strategy, and times to take a break. She encouraged me in my travels. Our many hours together on skis, bikes, and foot kept me focused on what this work is ultimately about. The two moose we skied past in winter, a couple of days before I submitted the final draft, were perfect.

There are no adequate ways to express my gratitude, Lisa. Without you, nothing.

NOTES

RESURGENCE

4 **A fox bit twin girls** "Mother's 'Nightmare' after Baby Twins Mauled by Fox," *BBC News*, June 7, 2010, https://www.bbc.com/news/10251349; "Fox Bites Off Baby's Finger," *The Guardian*, February 9, 2013, https://www.theguardian.com/uk/2013/feb/09/fox-bites-baby-finger; "Mum Describes the Terrifying Moment a Fox Attacked Her Baby," PlymouthLive, February 14, 2018, https://www.plymouthherald.co.uk/news/mum-describes-terrifying-moment-fox-1216052.

5 **Wildlife populations have declined 20 percent** The statistics in this paragraph are from the 2019 *Intergovernmental Panel on Biodiversity and Ecosystem Services Report (IPBES 2019)*, the *Living Planet Report 2020*, the *IUCN Red List 2021*, and "Decline of the North American Avifauna" (Sauer et al., 2019).

5 **The collapse is almost certain to get worse** From the 2019 *Intergovernmental Panel on Climate Change's Special Report on Climate Change and Land.*

5 **the weight of the world's mammalian life** "The Biomass Distribution on Earth" (Bar-On, Phillips, and Milo, 2018).

5 **Climate change is adding to the problems** *Climate Change 2021: The Physical Science Basis. Contribution of Working Group I to the Sixth Assessment Report of the Intergovernmental Panel on Climate Change*; "Global Warming Transforms Coral Reef Assemblages" (Hughes et al., 2018); *IPBES 2019*; and *Intergovernmental Panel on Climate Change Special Report on Global Warming of 1.5 Degrees C.*

6 **some animal populations are on the rebound** The humpback whale estimate is from Chris Wilkinson, "Humpback Whales Make Stunning Comeback in Southern Africa," *National Geographic*, July 18, 2019, https://www.nationalgeographic

.com/animals/article/humpback-whales-recovery-south-africa. The black bear estimates are from the California Department of Fish and Wildlife. Stork births were reported by the Knepp Castle Estate. Wolves in Europe are estimated in "Recovery of Large Carnivores in Europe's Modern Human-Dominated Landscapes" (Chapron et al., 2014).

FARMLAND VILLAINS

11 **By 1869, the last Dutch wolf was dead** This timeline appears in an article documenting the Luttlegeest wolf coauthored by twenty-eight European wildlife specialists. "The First Wolf Found in the Netherlands in 150 Years Was the Victim of a Wildlife Crime" (Gravendeel et al., 2013).

11 **"the very incarnation of destruction"** Quoted in *Of Wolves and Men* (Lopez, 1979), 175.

12 **An early-morning driver found the body** "Animal Killed by a Car Was 98% Certain a Wolf, Say Experts," *DutchNews.NL*, July 8, 2013, https://www.dutchnews.nl/news/2013/07/animal_killed_by_a_car_was_98.

14 **The wolf hadn't been killed by a car at all** "The First Wolf Found in the Netherlands" (Gravendeel et al., 2013). Jansman told me there was frustration in the wildlife biology community about how public relations were handled during the whole saga.

15 **prompting a headline** "Terrifying Footage of Wolf Prowling City Streets Looking for Its Next Meal," *Daily Mirror*, March 13 (Corcoran, 2015), https://www.mirror.co.uk/news/weird-news/terrifying-footage-wolf-prowling-city-5327412.

15 **Dutch hunters and government agencies** "Thousands of Wild Boar and Deer to Be Shot This Year in Veluwe," *NL Times*, July 9, 2021, https://nltimes.nl/2021/07/09/thousands-wild-boar-deer-shot-year-veluwe.

16 **Females are sexually receptive** The account of wolf biology below is drawn from Lopez (1979), Foresman (2012), and conversations with Diane Boyd.

17 **a system animal behaviorist Carl Safina calls *extended childcare*** *Beyond Words: What Animals Think and Feel* (Safina, 2015).

19 **She says this ranks her among the most fortunate** "The Woman Who Runs with the Wolves," *Sports Illustrated*, October 18 (Knize, 1993), https://vault.si.com/vault/1993/10/18/the-woman-who-runs-with-the-wolves-biologist-diane-boyd-has-spent-14-years-studying-the-animals-in-the-far-reaches-of-the-rockies.

21 **the most obvious place for a government-sponsored return** The story of the reintroduction and its ecological effects is documented in *Yellowstone Wolves: Science and Discovery in the World's First National Park* (D. W. Smith, Stahler, and MacNulty, 2020).

24 **A giant white truffle** "The Most Expensive Truffles in the World," Most Expensive Hub, https://mostexpensivehub.com/top-10-most-expensive-truffles-in-the-world.

25 **The scientific consensus** It has been a few years since a comprehensive study of the wolf population in Italy has been conducted. In 2015, a population range of up to eighteen hundred was described as "possibly underestimated." "One, No One, or One Hundred Thousand" (Galaverni et al., 2015), 13.

25 **The Apennine wolf feeds mainly** "Changes of Wolf (*Canis lupus*) Diet in Italy in Relation to the Increase of Wild Ungulate Abundance" (Meriggi et al., 2011).

26 **Wolves, they say, make ten thousand attacks** "France's Wolves Are Back. Now, Can It Protect Its Farmers?," *Christian Science Monitor*, April 4 (Davidson, 2018), https://www.csmonitor.com/World/Europe/2018/0404/France-s-wolves-are-back.-Now-can-it-protect-its-farmers.

27 **Wolves from Slovenia's Dinaric Alps have bred** "Long-Distance Dispersal Connects Dinaric-Balkan and Alpine Grey Wolf (*Canis lupus*) Populations" (Ražen et al., 2016).

27 **A pack of three adults and five pups** "Wolves Kill 120 Sheep near Dillon, Montana," August 28 (*Spokesman-Review*, 2009), https://www.spokesman.com/stories/2009/aug/28/wolves-kill-120-sheep-near-dillon-mont.

28 **The roots of the word *wilderness*** The story of the origin of the idea of wilderness in Western environmental thinking is told by Roderick Nash in *Wilderness and the American Mind* (Nash, 1967).

29 **the malaise of the industrial life George Monbiot calls *ecological boredom*** Monbiot uses this term in *Feral: Searching for Enchantment on the Frontiers of Rewilding* (Monbiot, 2013).

29 **Guillaume Chapron, a wildlife biologist** "Recovery of Large Carnivores in Europe's Modern Human-Dominated Landscapes" (Chapron et al., 2014).

29 **There are some striking examples of the coexistence model** "Rethinking the Study of Human-Wildlife Coexistence" (Pooley, Bhatia, and Vasava, 2021); "Coexistence of African Lions, Livestock, and People in a Landscape with Variable Human Land Use and Seasonal Movements" (Schuette, Creel, and Christianson, 2013); "Coexistence between Wildlife and Humans at Fine Spatial Scales" (Carter et al., 2012).

31 **Willem van Toorn, a well-known Dutch poet** Van Toorn's works are published in Dutch but have been written about in English in this context by Martin Drenthen. For example, "Reading Ourselves through the Land: Hermeneutics and the Ethics of Place" (Drenthen, 2011).

31 **turned dozens of military bases into stepping-stones** "Military Training Areas Facilitate the Recolonization of Wolves in Germany" (Reinhardt et al., 2019).

39 **McIntyre is an award-winning author** McIntyre's books include *The Rise of Wolf 8: Witnessing the Triumph of Yellowstone's Underdog* (McIntyre, 2019), *The Reign of Wolf 21: The Saga of Yellowstone's Legendary Druid Pack* (McIntyre, 2020), and *Redemption of Wolf 302: From Renegade to Yellowstone Alpha Male* (McIntyre, 2021).

40 **Roe deer were almost extinct in the Netherlands** The available data is a decade old, making it likely the numbers are higher today. "Roe Deer Population and Harvest Changes in Europe" (Burbaitė and Csányi, 2009); "Red Deer Population and Harvest Changes in Europe" (Burbaitė and Csányi, 2012); *Wildlife Comeback in Europe: The Recovery of Selected Mammal and Bird Species* (Deinet et al., 2013).

41 **Wild boar numbers are also booming** The clearinghouse for data on wild boar in Europe is at Enetwild, https://enetwild.com. Part of the reason to monitor wild boar is a concern about their ability to spread pathogens such as African swine fever.

42 **A landscape is no longer just scenery when wolves are around** This idea was expressed slightly differently by Kris Tompkins in a *Sierra Magazine* article about her work in wildlife conservation. Jason Mark, "The Fashion Executives Who Saved a Patagonia Paradise," Sierra, September 4, 2019, https://www.sierraclub.org/sierra/2019-5-september-october/feature/fashion-executives-who-saved-patagonian-paradise-doug-kris-tompkins.

42 **There are now three packs in the Veluwe** Updates on the status of Dutch wolves can be found on the website maintained by Wolven in Nederland at https://www.wolveninnederland.nl.

44 **They went from villainous beasts to loyal companions.** A compelling illustration of this transition is described in "Retriever of Souls" (Irvine and McHarg, 2021), especially 58–59.

46 **A farmer shared a video of the wolf** "Wolf Filmed Attacking Sheep in Noord Brabant; Action Is Needed, Farmers Say" (DutchNews.NL, 2020), https://www.dutchnews.nl/news/2020/05/wolf-filmed-attacking-sheep-in-noord-brabant-action-is-needed-farmers-say.

48 **In January 2018, a wolf wearing a radio collar** "First Confirmed Wolf in Belgium in a Hundred Years" (Huisman, 2018), https://wilderness-society.org/first-confirmed-wolf-belgium-100-years.

48 **But in June, Naya disappeared** "Belgium's First Sighted Wolf in a Century Feared Killed by Hunters" (*The Guardian*, 2019), https://www.theguardian.com/environment/2019/oct/02/belgiums-first-sighted-wolf-in-a-century-feared-killed-by-hunters.

49 **On New Year's Eve, the *Brussels Times* reported** Maïthé Chini, "New Female Wolf 'Noëlla' Spotted in Flanders," *Brussels Times*, December 31, 2019, https://www.brusselstimes.com/news/belgium-all-news/86758/new-female-wolf-noella-spotted-in-flanders-august-naya-christmas.

PRAIRIE PURITANS

53 **calls to mind two historic tragedies** One of the many versions of this story is told in *Great Plains Bison* (O'Brien, 2017).

53 **The sixty million bison the settlers found** Nobody knows precisely how many bison lived in North America before Europeans arrived. Common estimates fall in the range of thirty million to sixty million. The U.S. Fish and Wildlife Service cites a range of thirty million to seventy-five million. James Shaw discusses a number of different historical counts ranging from four million to "considerably more than 100 million" (Shaw, 1995). Some historians suggest bison numbers became elevated after diseases introduced by white settlers decimated the Indigenous people who used to hunt them.

53 **the animal they call the *iinnii* or buffalo** The words *buffalo* and *bison* refer to the same animal, but *buffalo* is the term preferred by the tribes.

54 **President Grant's Secretary of the Interior** Columbus Delano wrote in his 1873 report to the president on how to turn Indians into farmers: ". . . I would not seriously regret the total disappearance of the buffalo from our western prairies, in its effect on the Indians, regarding it rather as a means of hastening their sense of dependence upon the products of the soil and their own labors. . . ." Columbus Delano, Report of the Secretary of the Interior, U.S. Department of the Interior, October 31, 1873, Wikisource, https://en.wikisource.org/wiki/Report_of_the_Secretary_of_the_Interior/1873.

55 **starred in a documentary about his ancestor** The documentary is called *In the Spirit of Atatiće* and is available on YouTube at https://www.youtube.com/watch?v=S1WvkSN8zDQ.

58 **Plains bison (*Bison bison bison*) are one of two subspecies** Many of the details that follow are from *Mammals of Montana* (Foresman, 2012).

59 **migrated across the Bering land bridge 195,000 years ago** "Fossil and genomic evidence constrains the timing of bison arrival in North America" (Froese, Stiller, Heintzmen, et al., 2017).

61 **Hornaday, like Roosevelt, was an explorer** A compelling account of Hornaday's approach to conservation together with his many faults can be found in *Beloved Beasts: Fighting for Life in an Age of Extinction* (Nijhuis, 2021).

61 **When he was criticized for his racism, Hornaday claimed** Nijhuis shows how the white supremacy displayed by people like Hornaday filtered into the founding ideals of the North American environmental movement.

62 **With a new set of advocates** Many of the details of the early figures and events in the history of U.S. bison conservation can be found at All About Bison at www.allaboutbison.com.

67 **Ernest Seton, a turn-of-the-century naturalist, wrote** Seton is quoted in *Lost Tracks: Buffalo National Park 1909–1939* (Brower, 2008).

68 **Michael Soulé, a founder of the discipline of conservation biology** "What Is Conservation Biology?" (Soulé, 1985), 731.

68 **"Species conservation is more than skin deep"** Quoted in "Conservation: The Genome of the American West" (Marris, 2009).

70 **"the largest nature reserve in the contiguous United States"** See American Prairie's mission statement at https://www.americanprairie.org/mission-and-values.

71 **as well as animals from at least one other commercial ranch** "The Origin of America's Public Bison Herds in the United States" (Wood, 2000), 161 (fig.1).

71 **nobody knows for sure whether some cattle genes didn't sneak in** A study published in 2022 using the latest techniques for reading DNA found that all the bison herds in North America, including the Yellowstone herd, contain some cattle genes. Of all the bison sampled, Yellowstone had the lowest levels of cattle genes. But Geddes had been proven right. "Genomic Evaluation of Hybridization in Historic and Modern North American Bison (*Bison bison*)" (Stroupe, Forgacs, Harris, et al., 2022).

71 **Montana Fish, Wildlife, and Parks recently released a report** In May 2021, new Republican Governor Greg Gianforte scrapped the bison restoration plan and removed the document from the Fish, Wildlife, and Parks website. At time of writing (December 2021), the plan could still be accessed at https://drive.google.com/file/d/1lzYqc5NF9w-hv_ujWnPMNLP_vgy5KjkV/view.

73 **These four-legged ecosystem engineers are a keystone species** Bison have been shown to create a green wave of fertility as they move across the prairie (Geremia et al., 2019). Several of the species discussed in this book are keystone species. These include the wolf, the bison, and the beaver. An explanation of the keystone concept can be found in "The Keystone-Species Concept in Ecology and Conservation Management and Policy Must Explicitly Consider the Complexity of Interactions in Natural Systems" (Mills, Soulé, and Doak, 1993).

74 **their potential to transmit disease** The disease the ranchers worry about most is brucellosis. This bacterial disease is common in Montana's elk population and can cause cattle to abort their fetuses. States known to have brucellosis-infected cattle can be prevented from selling or moving their animals without expensive testing. About 60 percent of adult females in the Yellowstone bison herd are known to carry brucellosis according to the National Park Service. There are, however, no documented cases of brucellosis transmission taking place from bison to cattle.

75 ***Antilocapra americana* is high on the list for the title** Some of these biological details are from *Mammals of Montana* (Foresman, 2012). Others came from conversation with Andrew Jakes and from the Montana Department of Fish, Wildlife, and Parks' page on pronghorn at https://fwp.mt.gov/conservation/wildlife-management/antelope.

76 **The discipline is defined** "A Fence Runs through It: A Call for Greater Attention to the Influence of Fences on Wildlife and Ecosystems" (Jakes et al., 2018), 311.

76 **Studies in Alberta** Andrew Jakes, personal communication.

77 **Other than Arctic caribou, only mule deer** "Longest Terrestrial Migrations and Movements around the World" (Joly et al., 2019).

78 **Valuable traits disappear through an evolutionary process** See the discussion in "Conservation Genetics and North American Bison (*Bison bison*)" (Hedrick, 2009).

80 **can you be sure you have got the right animal back?** Emma Marris, knowing the high percentage of bison that possess cattle genes, puts it this way: "Are most bison really not bison?" (Marris, 2009).

82 **restoring natural grazing in protected areas** Natural grazing, particularly if it occurs in a patchy fashion, can enhance biodiversity by creating ecologically rich edges bordering different ecotypes. "Rewilding with Large Herbivores: Direct Effects and Edge Effects of Grazing Refuges on Plant and Invertebrate Communities" (van Klink, Ruifrok, and Smit, 2016).

82 **In most parts of Europe and North America, the landscape can support** "Exploring a Natural Baseline for Large-Herbivore Biomass in Ecological Restoration" (Fløjgaard et al., 2021).

82 **Julius Caesar wrote** *Caesar's Gallic War* (Caesar, 1869), bk. 6, ch 28, https://www.perseus.tufts.edu/hopper/text?doc=Perseus%3Atext%3A1999.02.0001%3Abook%3D6%3Achapter%3D28.

86 **the European bison's genome (*Bison bonasus*) shows** "Early Cave Art and Ancient DNA Record the Origin of European Bison" (Soubrier et al., 2016). For a discussion of this paper's findings, see "Mysterious Origin of European Bison Revealed Using DNA and Cave Art" (Marris, 2016).

89 **The Initiative brings together** Details of the work can be found at the Blackfeet Nation site at https://blackfeetnation.com/iinnii-buffalo-spirit-center.

90 **a marginal reduction in weight** Reported in Marris (2009).

RIVER ENGINEERS

101 **They built an average of one large dam** "The Anthropocene: Are Humans Now Overwhelming the Great Forces of Nature?" (Steffen, Crutzen, and McNeill, 2007). Former Secretary of the Interior Bruce Babbitt is also given credit for this observation (Klein, 1999). Babbitt was the first cabinet secretary to seriously advocate dam removals. He presided over the destruction of several dams during his tenure at the Department of Interior during the Clinton administration.

102 **Manifest destiny, according to influential nineteenth-century journalist John O'Sullivan** "Annexation" (O'Sullivan, 1845), https://pdcrodas.webs.ull.es/anglo/OSullivanAnnexation.pdf.

103 **The combination of dams and reservoirs can kill** "Reservoir Entrapment and Dam Passage Mortality of Juvenile Chinook Salmon in the Middle Fork Willamette River" (Keefer et al., 2012).

103 **A comprehensive 2020 report found migratory fish** "The Living Planet Index (LPI) for Migratory Freshwater Fish—Technical Report" (Deinet et al., 2020).

103 **The Columbia River used to see sixteen million salmon each year** "Salmon at River's End: The Role of the Estuary in the Decline and Recovery of Columbia River Salmon" (Bottom et al., 2005). The Klamath river numbers can be found in Stephen Most, "The Klamath," *Oregon Encyclopedia*, https://www.oregonencyclopedia.org/articles/klamath_river/#.YYlhk2DMI2x.

103 **Close to 30 percent of the Pacific salmon runs south of British Columbia** "Pacific Salmon Extinctions: Quantifying Lost and Remaining Diversity" (Gustafson et al., 2007).

105 **The Elwha Dam stood tall** The story of the building of the two Elwha dams and their life up to the time of the decision to remove them is told in detail in a history conducted by the U.S. National Park Service, "An Interpretive History of the Elwha River Valley and the Legacy of Hydropower on Washington's Olympic Peninsula" (Sadin, Vogel, and Miller, 2011).

105 **She told her grandchildren she watched salmon** The quotes in this paragraph and several following paragraphs are drawn from Linda Mapes's account of the dam removal, *Elwha: A River Reborn* (Mapes and Ringman, 2013).

106 **In 1986, the Lower Elwha Klallam filed a motion** An account of the Elwha tribe's role in initiating the dam removal process can be found in "Tribal Advocacy and the Art of Dam Removal: The Lower Elwha Klallam and the Elwha Dams" (Guarino, 2013).

106 **Tribal elder Bea Charles** Quoted in *Elwha: A River Reborn* (Mapes and Ringman, 2013), 114.

107 **Within weeks, steelhead trout spawned above the former dam site** The timing of the return of eight anadromous fish species above the former dam sites is detailed in "Reconnecting the Elwha River: Spatial Patterns of Fish Response to Dam Removal" (Duda et al., 2021).

112 **within two and a half years, nine runs of anadromous fish** "Reconnecting the Elwha River" (Duda et al., 2021).

112 **occasionally placed fish in various tributaries** "Relocation and Recolonization of Coho Salmon in Two Tributaries to the Elwha River: Implications for Management and Monitoring" (Liermann et al., 2017).

112 **Tons of sediment flushed out** "Large-Scale Dam Removal on the Elwha River, Washington, USA: Coastal Geomorphic Change" (Gelfenbaum et al., 2015).

113 **Juvenile coho heading to the ocean** "Large Coho Salmon Smolts Found Exiting Lake Sutherland," *Peninsula Daily News*, September 6 (*Peninsula Daily News*, 2019), https://www.peninsuladailynews.com/news/large-coho-salmon-smolts-found-exiting-lake-sutherland.

113 **A large salmon contains** "Why Fish Need Trees and Trees Need Fish," *Alaska Fish & Wildlife News*, Alaska Department of Fish and Game (Post, 2008), https://www.adfg.alaska.gov/index.cfm?adfg=wildlifenews.view_article&articles_id=407.

113 **bears, eagles, and raccoons grab** Many of the ecological relationships between salmon and their rainforest ecosystem presented in the next few paragraphs are detailed in *Pacific Salmon and Wildlife: Ecological Contexts, Relationships, and Implications for Management* (Cederholm et al., 2000) and in "Why Fish Need Trees and Trees Need Fish" (Post, 2008).

113 **The 137 species that directly benefit from salmon** Some sources claim more, but this number is cited in *Pacific Salmon and Wildlife* (Cederholm et al., 2000).

114 **Dippers on the Elwha have become bigger** "The Rapid Return of Marine-Derived Nutrients to a Freshwater Food Web Following Dam Removal" (Tonra et al., 2015).

118 **Humans aside, it is unlikely any species shaped the landscapes** Ben Goldfarb's book, *Eager: The Surprising, Secret Life of Beavers and Why They Matter*, has been at the forefront of an explosion of interest in the ecological value of beavers. The sections on beaver biology and ecology that follow draw heavily from his work (Goldfarb, 2019).

119 **Ernest Thompson Seton estimated as many as 400 million** *Lives of Game Animals* (Seton, 1929, quoted in Goldfarb, 2019), 448.

119 **By 1841, a trapper named Osborne Russell suggested** *Journal of a Trapper* (Russell, 1965; quoted in Goldfarb, 2019).

120 **As many as seventy-five times more ducks** "The Importance of Beaver to Wetland Habitats and Waterfowl in Wyoming" (McKinstry, Caffrey, and Anderson, 2001).

120 **The landscape—once so full of ponds** *Once They Were Hats: In Search of the Mighty Beaver* (Backhouse, 2015), 3.

124 **Beavers, like wolves, don't appear to be fountains of fertility** Some of these biological details are from *Mammals of Montana* (Foresman, 2012). Others are from *Eager* (Goldfarb, 2019) and from conversations with Ben Goldfarb.

125 **the complex ecological tapestry that beavers maintained** A good deal of what Mills observed has been confirmed by subsequent studies. "Beaver: Nature's Ecosystem Engineers" (Brazier et al., 2021).

125 **"May his tribe increase!"** *In Beaver World* (Mills, 1913; quoted in Goldfarb, 2019), 221.

126 **An Alberta study found beavers** "Wolves, White-Tailed Deer, and Beaver: Implications of Seasonal Prey Switching for Woodland Caribou Declines" (Latham et al., 2013).

126 **The U.S. government's Wildlife Services division kills** Reports are published annually on the U.S. Department of Agriculture's Wildlife Services website. The number 25,000 is the total in the USDA-APHIS 2020 report. Wildlife Services, Animal and Plant Health Inspection Service (APHIS), U.S. Department of Agriculture (USDA), https://www.aphis.usda.gov/aphis/ourfocus/wildlifedamage/pdr/?file=PDR-G_Report&p=2020:INDEX:.

127 **Whipple's studies found beaver dams** "Riparian Resilience in the Face of Interacting Disturbances: Understanding Complex Interactions between Wildfire, Erosion, and Beaver (*Castor canadensis*) in Grazed Dryland Riparian Systems of Low Order Streams in North Central Washington State, USA," Master's thesis, Eastern Washington University (Whipple, 2019).

127 **Wildland firefighters have valuable fire breaks** "Smokey the Beaver: Beaver-Dammed Riparian Corridors Stay Green during Wildfire throughout the Western United States" (Fairfax and Whittle, 2020).

127 **beavers reduced the flow of water cascading out of the hills** *River Otter Beaver Trial Science and Evidence Report* (Brazier et al., 2020), 70.

128 **"Castorids and salmonids possess millions of years"** *Eager* (Goldfarb, 2019), 113.

128 **Trout in the UK grew bigger in beaver-modified waterways** "The Response of a Brown Trout (*Salmo trutta*) Population to Reintroduced Eurasian Beaver (*Castor fiber*) Habitat Modification" (Needham et al., 2021).

128 **Juvenile chinook salmon in the Skagit estuary in Washington** "Beaver in Tidal Marshes: Dam Effects on Low-Tide Channel Pools and Fish Use of Estuarine Habitat" (Hood, 2012).

130 **signed an agreement with the U.S. Fish and Wildlife Service** The agreement signed when the current utility owners bought the dam is explicit, specifying that "no later than 25 years after the purchase date" the city and the utilities shall establish a program that "adequately and equitably protects, mitigates damage to, and enhances fish and wildlife resources . . . affected by the Eklutna Project." The agreement is available at "Fish and Wildlife Agreement: Snettisham and Eklutna Projects," Eklutna hydro.com, https://www.eklutnahydro.com/wp-content/uploads/2019/05/1991-Fish-and-Wildlife-Agreement.pdf.

131 **"Eklutna has given a lot to Anchorage's growth"** Meiklejohn is quoted in "Eklutna River Restoration Efforts Moving Forward," *Chugiak-Eagle River Star*, September 5 (Swann, 2017), https://www.alaskastar.com/2017-08-10/eklutna-river-restoration-efforts-moving-forward.

132 **John McPhee called shad** *The Founding Fish* (McPhee, 2002).

132 **Henry David Thoreau mourned for the declining fish** *A Week on the Concord and Merrimack Rivers* (Thoreau, 1980), 26.

133 **Pennsylvania alone has removed** American Rivers maintains a database of dam removals across the United States. The tally at the start of 2022 was nearing eighteen hundred dams. The database is available at "Map of U.S. Dams Removed since 1912," American Rivers, https://www.americanrivers.org/threats-solutions/restoring-damaged-rivers/dam-removal-map.

133 **Reductions in runoff from farms and cities bordering the Chesapeake Bay** The Chesapeake Bay Foundation runs a monitoring program for seagrass beds. See

"Bay Grasses," Chesapeake Bay Foundation, https://www.cbf.org/about-the-bay/more-than-just-the-bay/chesapeake-plants/bay-grasses.html.

133 **Juvenile American shad in the Potomac climbed** "The Return of American Shad to the Potomac River: 20 Years of Restoration," Final Report, Interstate Commission on the Potomac River Basin (Cummins, 2016).

134 **A 2020 report by the Atlantic States Marine Fisheries Commission** "Atlantic States Marine Fisheries Commission Annual Report 2020" (ASMFC, 2020).

134 **An early English visitor called it** *Travels and Works of Captain John Smith* (Smith, 1910).

135 **"Dolphins are swimming in the Potomac"** Latisha Johnson, "Breakfast Links: Dolphins Are Swimming in the Potomac. Are We Next?," Greater Greater Washington, October 3, 2019, https://ggwash.org/view/74133/breakfast-links-dolphins-are-swimming-and-mating-in-the-potomac. See also Karen Brulliard, "Dolphins Are Swimming, Mating, and Even Giving Birth in the Potomac," *Washington Post*, October 1 (Brulliard, 2019).

136 **a pretty good surrogate for a beaver dam** Not only are beaver dam analogues (BDAs) good surrogates for beaver dams; they can also help beavers build better dams where woody material is in short supply. Studies in Oregon showed BDAs substantially improve habitat for steelhead trout and increased the number of beavers in sections of the creek. "Ecosystem Experiment Reveals Benefits of Natural and Simulated Beaver Dams to a Threatened Population of Steelhead (*Oncorhynchus mykiss*)" (Bouwes et al., 2016).

138 **more than they contributed to carbon storage** Studies in Rocky Mountain National Park show that beaver habitat holds more carbon than habitat without beavers. "Landscape-Scale Carbon Storage Associated with Beaver Dams" (Wohl, 2013).

142 **The genetic differences that accrue** *Salmon: A Fish, the Earth, and the History of Their Common Fate* (Kurlansky, 2020), 54.

143 **when it comes to fish returning from the ocean** "Wild Chinook Salmon Survive Better Than Hatchery Salmon in a Period of Poor Production" (Beamish et al., 2012); "Reproductive Success in Wild and Hatchery Male Coho Salmon" (Neff et al., 2015).

143 **Increasing evidence suggests hatcheries slowly degrade** Some of this evidence came to my attention during an email exchange with Kyle Shedd of the Alaska Department of Fish and Game. Shedd shared preliminary results from an ongoing study of the fitness of hatchery fish compared to natural-origin fish at an Alaska Marine Science Symposium in January 2021. Kyle Shedd, "Reduced Relative Fitness in Stray Hatchery-Origin Pink Salmon from Large-Scale Segregated Hatcheries in Prince William Sound Alaska," Alaska Marine Science Symposium 2021. The findings suggest hatchery fish returning to streams outside their hatchery origin have about 50 percent of the reproductive success of natural-origin fish. If they breed with wild fish, they will degrade their genome.

143 **Hatchery fish tend to be smaller than wild fish** "Comparison of Life History Traits between First-Generation Hatchery and Wild Upper Yakima River Spring Chinook Salmon" (Knudsen et al., 2006).

144 **A *Los Angeles Times* opinion piece** "To Save Klamath River Salmon, Shut Down the Hatcheries," *Los Angeles Times*, June 13 (Leslie, 2019), https://www.latimes.com/opinion/op-ed/la-oe-leslie-salmon-hatcheries-klamath-river-20190613-story.html.

144 **from a twenty-five-year study** "The Road to Extinction Is Paved with Good Intentions: Negative Association of Fish Hatcheries with Threatened Salmon" (Levin, Zabel, and Williams, 2001).

FOREST MANAGERS

149 **Northern spotted owls (*Strix occidentalis caurina*) are chestnut-colored** "Scientific Evaluation of the Status of the Northern Spotted Owl," report prepared for the U.S. Fish and Wildlife Service (Courtney et al., 2004).

150 **are highly dependent on the cavities and broken tops** "Characteristics of Forests at Spotted Owl Nest Sites in the Pacific Northwest" (Hershey, Meslow, and Ramsey, 1998).

150 **a new Forest Service policy reduced logging** The "Northwest Forest Plan" set the rules for nearly twenty-five million acres of federal land in the Northwest. It is a multidecadal plan and is still in effect today. It can be found at Regional Ecosystem Office, "Northwest Forest Plan," U.S. Forest Service, https://www.fs.fed.us/r6/reo/overview.php.

152 **described the species as "circling the drain"** The biologist, Dominick DellaSala of the National Center for Conservation Science & Policy, was talking to Craig Welsh in an article for *Smithsonian Magazine* (Welch, 2009), https://www.smithsonianmag.com/science-nature/the-spotted-owls-new-nemesis-131610387.

153 **There were now more than 3.5 million of them** A respected source for U.S. bird numbers is the Partners in Flight, Population Estimates Database, available at https://pif.birdconservancy.org/population-estimate-database-scores.

155 **There is even a case in North Carolina** "Was an Owl the Real Culprit in the Peterson Murder Mystery?," *Audubon*, November 12 (Bargmann, 2016). I met with Kate Davis, a raptor specialist who had been called as a potential expert witness in the retrial of the person convicted in the murder. She took clay casts of a barred owl's talons to use as evidence if the case went to trial. Davis is convinced the victim was killed by a barred owl as she approached her house.

157 **in today's lingo, "conservation reliant"** "Recovery of Imperiled Species under the Endangered Species Act: The Need for a New Approach" (Scott et al., 2005).

157 **he acknowledged there were challenges but said the barred owl case was clear** While there is debate about the cause of the barred owl's range expansion, there is

no doubt about its reality nor its effects on spotted owls. See "Range Expansion of Barred Owls, Part I: Chronology and Distribution" (Livezey, 2009a) and "The Invasion of Barred Owls and Its Potential Effect on the Spotted Owl: A Conservation Conundrum" (Gutiérrez et al., 2007).

158 **because the Great Plains formed a barrier** "Range Expansion of Barred Owls, Part II: Facilitating Ecological Changes" (Livezey, 2009b).

158 **The alternative explanation for their migration** Wiens is not alone in supporting the connection to human expansion (Livezey, 2009b).

159 **Wiens's report, filed six months after the experiment finished** "Invader Removal Triggers Competitive Release in a Threatened Avian Predator" (Wiens et al., 2021).

159 **In the areas Hunt staked out in Washington** "Effects of Barred Owl (*Strix varia*) Removal on Population Demography of Northern Spotted Owls (*Strix occidentalis caurina*) in Washington and Oregon," 2019 Annual Report Prepared in Cooperation with the U.S. Fish and Wildlife Service, Bureau of Land Management (Wiens et al., 2020), 8.

160 **Spotted owls tend to use the canopy** "Three-Dimensional Partitioning of Resources by Congeneric Forest Predators with Recent Sympatry" (Jenkins et al., 2019).

160 **The government in British Columbia** British Columbia already has a captive breeding facility for spotted owls. All the remaining spotted owls may soon be brought into this facility for captive breeding, similar to what was done for the California Condor (Welch, 2009).

161 **received permits to kill sea lions** "US Allows Killing of Hundreds of Sea Lions to Save Struggling Salmon," *The Guardian*, August 15 (AP, 2020), https://www.theguardian.com/us-news/2020/aug/15/us-sea-lions-columbia-river-salmon-steelhead-trout.

161 **Emma Marris, author of *Wild Souls*** *Wild Souls: Freedom and Flourishing in the Non-Human World* (Marris, 2021). The "blood on our hands" quote comes from "Handle with Care," an essay in *Orion Magazine* (Marris, 2015), https://orionmagazine.org/article/handle-with-care.

164 ***Bison bonasus*, like *Bison bison*, is the largest terrestrial animal** *European Bison: The Nature Monograph* (Krasińska and Krasiński, 2013).

165 **The story that circulated in the media** "Wild Bison to Return to UK for First Time in 6,000 Years," *The Guardian*, July 10 (Carrington, 2020), https://www.theguardian.com/environment/2020/jul/10/wild-bison-to-return-to-uk-kent.

166 **European bison evolved after steppe bison bred with aurochs** "Early Cave Art and Ancient DNA Reveal the Origin of European Bison" (Soubrier et al., 2016).

168 **A newly developed Biodiversity Intactness Index** The Biodiversity Intactness Index is a project of the Natural History Museum in London. It can be accessed at https://www.nhm.ac.uk/our-science/our-work/biodiversity/predicts.html.

171 **Wildwood had recently garnered international headlines** "Wildcats Could Return to England after 200 Years" (BBC, 2021), https://www.bbc.com/news/uk-england-kent-56847548.

172 **The Royal Zoological Society of Scotland ran a study** "Distinguishing the Victim from the Threat: SNP-Based Methods Reveal the Extent of Introgressive Hybridization between Wildcats and Domestic Cats in Scotland and Inform Future *in situ* and *ex situ* Management Options for Species Restoration" (Senn et al., 2019).

172 **The study spurred a report the following year** "Conservation of the Wildcat (*Felis silvestris*) in "Scotland: Review of the Conservation Status and Assessment of Conservation Activities" (Breitenmoser, Lanz, and Breitenmoser-Würsten, 2019).

172 **Wildwood joined two other conservation organizations** "A Preliminary Feasibility Assessment for the Reintroduction of the European Wildcat to England and Wales" (MacPhearson et al., 2019).

175 **site of the UK's first beaver reintroduction** "The Ham Fen Beaver Project: First Report to the Peoples Trust for Endangered Species," Wildlife Conservation Research Unit, University of Oxford (Campbell and Tattersall, 2003).

176 **In his book *Wild Ones*** *Wild Ones: A Sometimes Dismaying, Weirdly Reassuring Story about Looking at People Looking at Animals in America* (Mooallem, 2013).

177 **Sweet chestnut coppice, much of it now overgrown** Some of these human uses are detailed in "The Blean: Canterbury and Swale's Ancient Woodlands" (Kent Wildlife Trust, 2011).

180 **I was struggling to keep pace with Mario Cipollone** Sections of this chapter are reprinted from "Why Italians Are Growing Apples for Wild Bears," © 2020, Christopher J. Preston, as first printed in *The Atlantic*.

181 **one of the rarest of the world's fourteen or so subspecies of brown bear** There is some debate about how many subspecies of bear there actually are. Estimates range from eight to sixteen. Jon Swenson, co-chair of the European Expert Brown Bear Team, told me that fourteen is a commonly accepted number. He added that the whole idea of subspecies is disappearing and being replaced by categorizing wildlife according to distinct populations. See *Bears of the World: Ecology, Conservation and Management* (Penteriani and Melleti, 2020).

182 **Italian zoologist Giuseppe Altobello classified the Marsican bear** *Fauna dell'Abruzzo e del Molise. Vertebrati, mammiferi. IV. I Carnivori (Carnivora)* (Altobello, 1921), http://www.storiadellafauna.com/wp-content/uploads/2020/03/Fauna-dellAbruzzo-e-del-Molise-nuove-forme-di-mammiferi-italiani-In-Molise-Rivista-regionale-illustrata-Anni-I-n.-4-agosto-dicembre-1923.pdf.

182 **a distinctive lower jaw** "Cranial Morphometrics of the Apennine Brown Bear (*Ursus arctos marsicanus*) and Preliminary Notes on the Relationships with Other Southern European Populations" (Loy et al., 2008).

183 **The Marsican bear is Italy's largest endemic carnivore** The brown bear (*Ursus arctos*) in Italy's Alps in the north tends to be a bit bigger than the Marsican bear. But this brown bear is reintroduced, and so some people suggest, technically speaking, it is no longer endemic. Many of the details about the Marsican bear below can be found in Salviamo l'Orso's fact sheet (Salviamo l'Orso, 2020).

187 **A series of reintroductions begun in the early 1990s** A comprehensive report on the reintroductions was published as *Chamois International Congress Proceedings: Lamadei Peligni, Majella National Park, Italy, 17–19 June 2014* (Antonucci and Di Domenico, 2015).

190 **When wildlife declined in the last century from overhunting and trapping** "Anthropogenic Drivers Leading to Population Decline and Genetic Preservation of the Eurasian Griffon Vulture (*Gyps fulvus*)" (Pirastru et al., 2021).

190 **slow but steady recovery through the release of captive-bred birds** "Status of the California Condor and Efforts to Achieve Its Recovery," report by the American Ornithologists' Union and Audubon (Walters et al., 2008).

OCEAN PARTNERS

202 **One Alaska fisherman reported a sperm whale** "Sperm Whales Awe and Vex Alaska Fishermen" (Woodford, 2003).

203 **Recently fishermen came up with something** "The Slinky Pot," *Pacific Fishing* (Mintz, 2021), https://www.pacificfishing.com/featured_stories/0321_story2.html.

204 **Fishermen report the slinky pots use less bait** "Off the Hook: Slinky Pots Revolutionize Alaska's Blackcod Fishery," *National Fisherman*, May 3 (Hagenbuch, 2021), https://www.nationalfisherman.com/boats-gear/off-the-hook-slinky-pots-revolutionize-alaska-s-blackcod-fishery.

204 **Whales are the largest animals in the history of life** Many of the details in the next few paragraphs come from Rebecca Giggs's wonderful book *Fathoms: The World in the Whale* (Giggs, 2020).

205 **One bowhead taken by Indigenous Americans** "Age and Growth Estimates of Bowhead Whales (*Balaena mysticetus*) via Aspartic Acid Racemization" (George et al., 2011).

205 **The effect of whales on marine ecosystems** "Whales as Marine Ecosystem Engineers" (Roman et al., 2014).

206 **Right whales and bowheads were soon almost gone** See also "A Review of Old Basque Whaling and Its Effect on the Right Whales (*Eubalaena glacialis*) of the North Atlantic" (Aguilar, 1986), which puts the expansion of Basque whaling into the North Atlantic even earlier.

206 **A particularly valuable oil** "Oil, Spermaceti, Ambergris, and Teeth: Products of the Nineteenth-Century Pacific Sperm-Whaling Industry" (Shoemaker, 2019).

206 **Australian author Rebecca Giggs quotes from** *Fathoms* (Giggs, 2020), 43.

207 **A humpback can have as many as eight hundred baleen plates** *Encyclopedia of Marine Mammals*, 3rd ed. (Würsig, Thewissen, and Kovacs, 2018).

209 **Millions of whales had been killed** These numbers are sprinkled around much of the literature on whales and their recovery. They can be found in Doughty et al. (2016), Pershing et al. (2010), Roman et al. (2014), and Roman and McCarthy (2010).

213 **Passengers on one vessel recently sailing off the Antarctic Peninsula** Philip Hoare, "Seeing 1,000 Glorious Fin Whales Back from Near Extinction Is a Rare Glimmer of Hope," *The Guardian*, January 17, 2022, https://www.theguardian.com/commentisfree/2022/jan/17/glorious-fin-whales-extinction-hope-antarctic-peninsular.

213 **Southern Ocean right whales had recovered from a low** "Southern Right Whale (*Eubalaena australis*) 5-Year Review: Summary and Evaluation" (Austin, 2021).

213 **Bowhead whales in the Bering, Chukchi, and Beaufort seas have tripled** "Bowhead Whale (*Balaena mysticetus*)," The IUCN Red List of Threatened Species, 2018, https://www.iucnredlist.org/species/2467/50347659.

213 **They remain critically endangered in Antarctica** "Towards Population-Level Conservation in the Critically Endangered Antarctic Blue Whale: The Number and Distribution of Their Populations" (Attard, Beheregaray, and Möller, 2016).

214 **Short-lived warming events like "The Blob"** "'The Blob' Invades Pacific, Flummoxing Climate Experts" (Kintisch, 2015).

222 **In 2015, a conservation biologist at the University of Vermont** "Global Nutrient Transport in a World of Giants" (Doughty et al., 2016).

222 **The dominant view of scientists** "Terrestrial Nutrient Cycling" (Chapin, Matson, and Mooney, 2002).

222 **"The past was a world of giants"** "Global Nutrient Transport in a World of Giants" (Doughty et al., 2016), 868.

223 **Blue whales in the North Pacific consume** "Baleen Whale Prey Consumption Based on High-Resolution Foraging Measurements" (Savoca et al., 2021).

224 **Iron concentration in whale fecal plumes** "Whales as Marine Ecosystems Engineers" (Roman et al., 2014), 381.

224 **Much of the lost weight is emitted** An account of the transfer of energy in fin whales can be found in "All Creatures Great and Smaller: A Study in Cetacean Life History Energetics" (Lockyer, 2007).

224 **Baleen whales alone ate twice the weight in krill** "Baleen Whale Prey Consumption Based on High-Resolution Foraging Measurements" (Savoca et al., 2021).

224 **blue whales might have moved 24,000 kilograms of nitrogen** "Whales as Marine Ecosystems Engineers" (Roman et al., 2014), 381.

224 **The snow is made up of** "The Importance of 'Marine Snow'" (Turley, 2002).

225 **"There is quite a lot of controversy about it"** Biuw coauthored a paper skeptical of the effect of whales on nutrient transport: "A Critical Evaluation of Whales as Ecosystem Engineers" (Wassman, Biuw, and Haug, 2021).

229 **In July 2012, a specially created entity** "Village Science Meets Global Discourse: The Haida Salmon Restoration Corporation's Ocean Iron Fertilization Experiment" (Buck, 2018).

230 **It turned out the controversial experiment in ocean fertilization** "Murky Waters: Ambiguous International Law for Ocean Fertilization and Other Geoengineering" (Wilson, 2013).

230 **Although it did create a phytoplankton bloom, marine ecologists** "Satellite Bio-optical and Altimeter Comparisons of Phytoplankton Blooms Induced by Natural and Artificial Iron Addition in the Gulf of Alaska" (Xiu, Thomas, and Chai, 2014).

230 **the contribution of sperm whales to carbon sequestration** "Iron Defecation by Sperm Whales Stimulates Carbon Export in the Southern Ocean" (Lavery et al., 2010).

231 **Roman looked at the nitrogen blue whales get from krill** "Whales as Marine Ecosystems Engineers" (Roman et al., 2014).

231 **A recent paper calculates that whale recovery in the Southern Ocean** "Baleen Whale Prey Consumption Based on High-Resolution Foraging Measurements" (Savoca et al., 2021).

232 **The food eaten by a blue whale** "Whales as Marine Ecosystems Engineers" (Roman et al., 2014).

232 **including some that evolved specifically for the carcasses of certain whales** "Whale-Fall Ecosystems: Recent Insights into Ecology, Paleoecology, and Evolution" (Smith et al., 2015).

232 **The carbon that sinks with a forty-ton gray whale** "Whales as Marine Ecosystems Engineers" (Roman et al., 2014).

232 **He and his collaborators looked at eight species of baleen whales** "The Impact of Whaling on the Ocean Carbon Cycle: Why Bigger Was Better" (Pershing et al., 2010).

232 **One estimate suggests $2 million worth of climate harm** "Nature's Solution to Climate Change" (Chami et al., 2019). There is some debate over whether the recent enthusiasm for whales as carbon-capture devices is being exaggerated. See, for example, "A Critical Evaluation of Whales as Ecosystem Engineers" (Wassman, Biuw, and Haug, 2021).

233 **liked a natural approach** Using animals to restore carbon sinks has been called *natural geoengineering.* "How 'Natural Geoengineering' Can Help Slow Global Warming" (Schmitz, 2016), https://e360.yale.edu/features/how_natural_geo-engineering_can_help_slow_global_warming.

233 **"gardeners in the greenhouse"** *Fathoms* (Giggs, 2020), 59.

237 **Sixty years ago, you would have been lucky to spot a sea otter** Many of the details in the next few paragraphs are drawn from *The Return of the Sea Otter: The Story of the Animal That Evaded Extinction on the Pacific Coast* (McLeish, 2018) and a collection of essays in the book *Sea Otter Conservation* (Larson, Bodkin, and VanBlaricom, 2015).

238 **When Vitus Bering's expedition limped home** *The Sea Otter* (Silverstein, Silverstein, and Silverstein, 1995), 35.

239 **Four sea otter hides could buy a house** Linda Nichol, "Conservation in Practice," in *Sea Otter Conservation* (Larson et al., 2015), 372.

239 **biologists decided to relocate otters from the Aleutian Islands** "History and Status of Translocated Sea Otter Populations in North America" (Jameson et al., 1982).

240 **the suspected culprit is killer whales** " Brenda Ballachey and James Bodkin, "Challenges to Sea Otter Recovery and Conservation," in *Sea Otter Conservation* (Larson, Bodkin, and VanBlaricom, 2015), 73–75.

241 **Some marine biologists think the killer whales might be to blame** "Complex Trophic Interactions in Kelp Forest Ecosystems" (Estes et al., 2004).

243 **Kelp is crucial to the health of the nearshore environment** *The Biology and Ecology of Giant Kelp Forests* (Schiel and Foster, 2015).

244 **A paper he wrote with John Palmisano on otters and kelp** "Sea Otters: Their Role in Structuring Nearshore Communities" (Estes and Palmisano, 1974).

245 **In 2012, he and four collaborators conducted a study** "Do Trophic Cascades Affect the Storage and Flux of Atmospheric Carbon? An Analysis of Sea Otters and Kelp Forests" (Wilmers et al., 2012). A different study looking at macroalgae outside of their connection to otters estimated they could sequester close to 200 million tonnes of carbon each year, equivalent to the emissions of 185 coal-fired power plants (Krause-Jensen et al., 2016).

247 **Macroinvertebrates, such as the shellfish so prized** "Behavioral Responses across a Mosaic of Ecosystem States Restructure a Sea Otter–Urchin Trophic Cascade" (Smith et al., 2021).

248 **Northwest coast tribes have a twelve-thousand-year history** This history is detailed in Anne K. Salomon, K̲ii'iljuus Barb J. Wilson, Xanius Elroy White, Nick Tanape Sr., and Tom Mexsis Happynook, "First Nations Perspectives on Sea Otter Conservation in British Columbia and Alaska: Insights into Coupled Human–Ocean Systems," in *Sea Otter Conservation* (Larson et al., 2015), 301–331. The chapter was produced as part of Coastal Voices, a project made up of "a diverse group of Indigenous leaders, knowledge holders, scientists and artists from British Columbia and Alaska working together, discussing and planning for the profound changes triggered by the return of sea otters." See http://coastalvoices.net.

249 **"Traditionally," says Tom Happynook of the Nuu-Cha-Nulth** "We had people who were responsible for taking care of the resources." Video interview with Tom Happynook as part of the Coastal Voices project, http://coastalvoices.net/tom16.

249 **that displayed what some have called a "cultivated abundance"** "An Indigenous Model of a Contested Pacific Herring Fishery in Sitka, Alaska" (Thornton and Kitka, 2015).

250 **"It was expected that you would be sustainable"** "Visioning the Future of Kelp Forest, Sea Otter, and Human Interactions: Workshop Summary Report" (Salomon, Burt, Wilson, et al., 2018), p.21.

250 **"You had better go to the outside waters"** The quote by Deborah Head is used with her permission and drawn from an interview conducted by Sonia Ibarra as part of the Apex Predators, Ecosystems, and Community Sustainability (APECS) project, at http://apecs-ak.org.

250 **"My process perpetuates an Indigenous protocol"** "Furs and Future" (Williams, 2021), https://magazine.firstalaskans.org/issue/spring-2021/furs-futures.

253 **The tribe filed suit** "Sitka Tribe Sues State, Claiming Mismanagement of the Herring Fishery" (Rose and Kwong, 2018), https://www.kcaw.org/2018/12/18/sitka-tribe-sues-state-claims-mismanagement-of-herring-fishery.

254 **Peter Williams sees the relationship with otters** *Return of the Sea Otter* (McLeish, 2018), 142.

255 **Beston also said this about animals** *The Outermost House* (Beston, 1962), 26.

CREATIVITY AND COURAGE

260 **Potawatomi plant ecologist Robin Kimmerer points out** *Braiding Sweetgrass: Indigenous Wisdom, Scientific Knowledge and the Teachings of Plants* (Kimmerer, 2020), 334

260 **The most recent assessment report on climate** The statistics that follow are from "Summary for Policymakers," *Climate Change 2021: The Physical Science Basis. Contribution of Working Group I to the Sixth Assessment Report of the Intergovernmental Panel on Climate Change* (Masson-Delmotte et al., 2021).

261 **United Nations Secretary-General António Guterres said the report** "IPCC Report: 'Code Red' for Human Driven Global Heating, Warns UN Chief," UN News, August 9, 2021, https://news.un.org/en/story/2021/08/1097362.

262 **Sir David Attenborough wrote what he called his "witness statement"** *A Life on Our Planet: My Witness Statement and a Vision for the Future* (Attenborough, 2020).

262 **Nature-based responses to climate change** "Natural Climate Solutions" (Griscom et al., 2017).

262 **It is not just what whales do for phytoplankton** "How 'Natural Geoengineering' Can Help Slow Global Warming" (Schmitz, 2016), https://e360.yale.edu/features/how_natural_geo-engineering_can_help_slow_global_warming. The idea animals can play a role in climate change solutions is sometimes called *animating the carbon cycle*.

267 **But here they were, swimming outside cement channels** I called the Clearwater Hatchery to ask about the likely origin of the salmon I saw above the fish trap. The hatchery specialist there told me he was not surprised there were a few salmon spawning in the creek outside of the hatchery. It was hard to know if they were strays from this year's return or if they lived independently of any hatchery production. He told me the Lewiston Dam, which was removed in the 1970s, had wiped out most of the chinook run in the Clearwater River and its tributaries (which included the Lochsa). His view was that all the chinook above the former dam site at Lewiston were descendants of hatchery fish reintroduced after the dam came down. The dam had a relatively ineffective fish ladder. This meant the passage of entirely wild fish through the fish ladder for the half century the dam stood would have been difficult. Perhaps it was difficult. But, important for my imagination, not impossible.

REFERENCES

Aguilar, Alex. 1986. "A Review of Old Basque Whaling and Its Effect on the Right Whales (*Eubalaena glacialis*) of the North Atlantic." *Report—International Whaling Commission* 10 (10): 191–199.

Altobello, G. 1921. *Fauna dell'Abruzzo e del Molise. Vertebrati, mammiferi. IV. I Carnivori (Carnivora).* Campobasso, Italy: Colitti. http://www.storiadellafauna.com/wp-content/uploads/2020/03/Fauna-dellAbruzzo-e-del-Molise-nuove-forme-di-mammiferi-italiani-In-Molise-Rivista-regionale-illustrata-Anni-I-n.-4-agosto-dicembre-1923.pdf.

Antonucci, A., and G. Di Domenico. 2015. *Chamois International Congress Proceedings: Lamadei Peligni, Masella National Park, Italy, 17–19 June 2014.* Pescari, Italy: Majambiente Edizioni.

AP. 2020. "U.S. Allows Killing of Hundreds of Sea Lions to Save Struggling Salmon." *The Guardian*, August 15, 2020.

Atlantic States Marine Fisheries Commission (ASMFC). 2020. "Atlantic States Marine Fisheries Commission Annual Report 2020." Arlington, VA, ASMFC.

Attard, Catherine R. M., Luciano B. Beheregaray, and Luciana M. Möller. 2016. "Towards Population-Level Conservation in the Critically Endangered Antarctic Blue Whale: The Number and Distribution of Their Populations." *Scientific Reports* 6 (1): 1–11. https://doi.org/10.1038/srep22291.

Attenborough, David. 2020. *A Life on Our Planet: My Witness Statement and a Vision for the Future.* London: Penguin.

Austin, Heather. 2021. "Southern Right Whale (*Eubalaena australis*) 5-Year Review: Summary and Evaluation." Office of Protected Services, U.S. National Marine Fisheries Service, National Oceanic and Atmospheric Administration, Silver Spring, MD.

Backhouse, Frances. 2015. *Once They Were Hats: In Search of the Mighty Beaver.* Toronto: ECW Press.

Bargmann, J. 2016. "Was an Owl the Real Culprit in the Peterson Murder Mystery?" *Audubon*, November.

Bar-On, Yinon M., Rob Phillips, and Ron Milo. 2018. "The Biomass Distribution on Earth." *Proceedings of the National Academy of Sciences* 115 (25): 6506–6511. https://doi.org/10.1073/PNAS.1711842115.

BBC. 2021. "Wildcats Could Return to England after 200 Years." BBC.com, April 22.

Beamish, R. J., R. M. Sweeting, C. M. Neville, K. L. Lange, T. D. Beacham, and D. Preikshot. 2012. "Wild Chinook Salmon Survive Better Than Hatchery Salmon in a Period of Poor Production." *Environmental Biology of Fishes* 94 (1): 135–148. https://doi.org/10.1007/S10641-011-9783-5/FIGURES/9.

Beston, Henry. 1962. *The Outermost House*. New York: Viking Compass.

Bottom, D. L., C. A. Simenstad, J. Burke, A. M. Baptista, D. A. Jay, K. K. Jones, E. Casillas, and M. H. Schiewe. 2005. "Salmon at River's End: The Role of the Estuary in the Decline and Recovery of Columbia River Salmon." NOAA Technical Memo NMFS-NWFSC-68, U.S. Department of Commerce, Portland, OR.

Bouwes, Nicolaas, Nicholas Weber, Chris E. Jordan, W. Carl Saunders, Ian A. Tattam, Carol Volk, Joseph M. Wheaton, and Michael M. Pollock. 2016. "Ecosystem Experiment Reveals Benefits of Natural and Simulated Beaver Dams to a Threatened Population of Steelhead (*Oncorhynchus mykiss*)." *Scientific Reports* 6 (1): 1–12. https://doi.org/10.1038/srep28581.

Brazier, R. E., M. Elliott, E. Andison, R. E. Auster, S. Bridgewater, P. Burgess, J. Chant, et al. 2020. *River Otter Beaver Trial Science and Evidence Report*. Exeter, UK: Devon Wildlife Trust.

Brazier, Richard E., Alan Puttock, Hugh A. Graham, Roger E. Auster, Kye H. Davies, and Chryssa M. L. Brown. 2021. "Beaver: Nature's Ecosystem Engineers." *WIRES Water* 8 (1). https://doi.org/10.1002/WAT2.1494.

Breitenmoser, U., T. Lanz, and C. Breitenmoser-Würsten. 2019. "Conservation of the Wildcat (*Felis silvestris*) in Scotland: Review of the Conservation Status and Assessment of Conservation Activities." Cat Specialist Group, Species Survival Commission (SSC), International Union for Conservation of Nature (IUCN).

Brower, Jennifer. 2008. *Lost Tracks: National Buffalo Park 1909–1939*. Edmonton, Canada: Athabasca University Press.

Brulliard, Karen. 2019. "Dolphins Are Swimming, Mating, and Even Giving Birth in the Potomac." *Washington Post*, October 1, 2019.

Brussels Times. 2019. "New Female Wolf 'Noëlla' Spotted in Flanders." December 31.

Buck, Holly Jean. 2018. "Village Science Meets Global Discourse: The Haida Salmon Restoration Corporation's Ocean Iron Fertilization Experiment." In *Geoengineering Our Climate? Ethics, Politics and Governance*, edited by J. Blackstock and S. Low, 314. London: Routledge.

Burbaitė, Lina, and Sándor Csányi. 2009. "Roe Deer Population and Harvest Changes in Europe." *Estonian Journal of Ecology* 58. https://doi.org/10.3176/eco.2009.3.02.

Burbaitė, Lina, and Sándor Csányi. 2012. "Red Deer Population and Harvest Changes in Europe." *Acta Zoologica Lituanica* 20 (4): 179–188. https://doi.org/10.2478/V10043-010-0038-Z.

Caesar, Julius. 1869. *Caesar's Gallic War*. Edited by W. A. McDevitte and W. S. Bohn. New York: Harper and Brothers. https://www.perseus.tufts.edu/hopper/text?doc=Perseus%3Atext%3A1999.02.0001%3Abook%3D6%3Achapter%3D28.

Campbell, R. D., and F. Tattersall. 2003. "The Ham Fen Beaver Project: First Report to the Peoples Trust for Endangered Species." Wildlife Conservation Research Unit, Department of Zoology, University of Oxford, Oxford.

Carrington, D. 2020. "Wild Bison to Return to the UK for the First Time in 6,000 Years." *The Guardian*, July 10, 2020.

Carter, Neil H., Binoj K. Shrestha, Jhamak B. Karki, Narendra Man Babu Pradhan, and Jianguo Liu. 2012. "Coexistence between Wildlife and Humans at Fine Spatial Scales." *Proceedings of the National Academy of Sciences* 109 (38): 15360–15365. https://doi.org/10.1073/PNAS.1210490109.

Cederholm, C. J., D. H. Johnson, R. E. Bilby, L. G. Dominguez, A. M. Garrett, W. H. Graeber, E. L. Greda, et al. 2000. *Pacific Salmon and Wildlife: Ecological Contexts, Relationships, and Implications for Management*. 2nd ed. Olympia, WA: Washington Department of Fish and Wildlife.

Chami, R., T. Cosimano, C. Fullenkamp, and S. Oztosun. 2019. "Nature's Solution to Climate Change." *Finance and Development* (International Monetary Fund) 56 (4) (December).

Chapin, F. Stuart, Pamela A. Matson, and Harold A. Mooney. 2002. "Terrestrial Nutrient Cycling." *Principles of Terrestrial Ecosystem Ecology*, 197–223. New York: Springer Verlag. https://doi.org/10.1007/0-387-21663-4_9.

Chapron, G., P. Kaczensky, J. D. Linnell, M. von Arx, D. Huber, H. Andrén, J. V. López-Bao, et al. 2014. "Recovery of Large Carnivores in Europe's Modern Human-Dominated Landscapes." *Science* 346 (6216): 1517–1519. https://doi.org/10.1126/SCIENCE.1257553.

Corcoran, Liam. 2015. "Terrifying Footage of Wolf Prowling City Streets Looking for Its Next Meal." *Daily Mirror*, March 13.

Courtney, S. P., J. A. Blakesley, R. E. Bigley, M. L. Cody, J. P. Dumbacher, R. C. Fleischer, A. Franklin, et al. 2004. "Scientific Evaluation of the Status of the Northern Spotted Owl." Sustainable Ecosystems Institute, Portland, OR.

Cummins, J. 2016. "The Return of American Shad to the Potomac River: 20 Years of Restoration." Final Report. Interstate Commission on the Potomac River Basin, Rockville, MD.

Davidson, C. 2018. "France's Wolves Are Back. Now, Can It Protect Its Farmers?" *Christian Science Monitor* (April).

Deinet, S., C. Ieronymidou, L. McRae, I. J. Burfield, R. P. Foppen, B. Collen, and M. Böhm. 2013. *Wildlife Comeback in Europe: The Recovery of Selected Mammal and Bird Species*. Final

Report to Rewilding Europe by ZSL, Birdlife International, and the European Bird Census Council. London: ZSL.

Deinet, S., K. Scott-Gatty, H. Rotton, W. M. Twardek, V. Marconi, L. McRae, L. J. Baumgartner, et al. 2020. "The Living Planet Index (LPI) for Migratory Freshwater Fish. Living Planet Index: Technical Report." Groningen, NL: World Fish Migration Foundation.

Doughty, Christopher E., Joe Roman, Søren Faurby, Adam Wolf, Alifa Haque, Elisabeth S. Bakker, Yadvinder Malhi, John B. Dunning, and Jens Christian Svenning. 2016. "Global Nutrient Transport in a World of Giants." *Proceedings of the National Academy of Sciences of the United States of America* 113 (4): 868–873. https://doi.org/10.1073/pnas.1502549112.

Drenthen, M. 2011. "Reading Ourselves through the Land: Landscape Hermenuetics and the Ethics of Place." In *Placing Nature on the Borders of Religion, Philosophy, and Ethics*, edited by Forrest Clingerman and M. Dixon, 123–138. Farnham, UK: Ashgate.

Duda, Jeffrey J., Christian E. Torgersen, Samuel J. Brenkman, Roger J. Peters, Kathryn T. Sutton, Heidi A. Connor, Phil Kennedy, et al. 2021. "Reconnecting the Elwha River: Spatial Patterns of Fish Response to Dam Removal." *Frontiers in Ecology and Evolution* 9 (December). https://doi.org/10.3389/fevo.2021.765488.

DutchNews.NL. 2020. "Wolf Filmed Attacking Sheep in Noord Brabant; Action Is Needed, Farmers Say." DutchNews.NL, May 23. https://www.dutchnews.nl/news/2020/05/wolf-filmed-attacking-sheep-in-noord-brabant-action-is-needed-farmers-say.

Estes, J. A., E. M. Danner, D. F. Doak, B. Konar, A. M. Springer, P. D. Steinberg, M. T. Tinker, and T. M. Williams. 2004. "Complex Trophic Interactions in Kelp Forest Ecosystems." *Bulletin of Marine Science* 74 (3): 621–638.

Estes, James A. J. A., and J. F. Palmisano. 1974. "Sea Otters: Their Role in Structuring Nearshore Communities." *Science* 185 (4156): 1058.

Fairfax, Emily, and Andrew Whittle. 2020. "Smokey the Beaver: Beaver-Dammed Riparian Corridors Stay Green during Wildfire throughout the Western United States." *Ecological Applications* 30 (8): e02225. https://doi.org/10.1002/EAP.2225.

Fløjgaard, Camilla, Pil Birkefeldt Møller Pedersen, Christopher J. Sandom, Jens-Christian Svenning, and Rasmus Ejrnæs. 2021. "Exploring a Natural Baseline for Large-Herbivore Biomass in Ecological Restoration." *Journal of Applied Ecology* 59 (1): 18–24. https://doi.org/10.1111/1365-2664.14047.

Foresman, Kerry Ryan. 2012. *Mammals of Montana*. 2nd ed. Missoula, MT: Mountain Press Publishing Company.

Froese, Duane, Mathias Stiller, Peter D. Heintzman, Alberto V. Reyes, Grant D. Zazula, André E. R. Soares, Matthias Meyer, et al. 2017. "Fossil and Genomic Evidence Constrains the Timing of Bison Arrival in North America." *Proceedings of the National Academy of Sciences* 114: 3457–3462.

Galaverni, Marco, Romolo Caniglia, Elena Fabbri, Pietro Milanesi, and Ettore Randi. 2015. "One, No One, or One Hundred Thousand: How Many Wolves Are There Currently in Italy?" *Mammal Research* 61 (1): 13–24. https://doi.org/10.1007/S13364-015-0247-8.

Gelfenbaum, Guy, Andrew W. Stevens, Ian Miller, Jonathan A. Warrick, Andrea S. Ogston, and Emily Eidam. 2015. "Large-Scale Dam Removal on the Elwha River, Washington, USA: Coastal Geomorphic Change." *Geomorphology* 246 (October): 649–668. https://doi.org/10.1016/J.GEOMORPH.2015.01.002.

George, John C, Jeffrey Bada, Judith Zeh, Laura Scott, Stephen E Brown, Todd O'Hara, and Robert Suydam. 2011. "Age and Growth Estimates of Bowhead Whales (*Balaena mysticetus*) via Aspartic Acid Racemization." *Canadian Journal of Zoology* 77 (4): 571–580. https://doi.org/10.1139/Z99-015.

Geremia, Chris, Jerod A. Merkle, Daniel R. Eacker, Rick L. Wallen, P. J. White, Mark Hebblewhite, and Matthew J. Kauffman. 2019. "Migrating Bison Engineer the Green Wave." *Proceedings of the National Academy of Sciences of the United States of America* 116 (51): 25707–25713. https://doi.org/10.1073/PNAS.1913783116/-/DCSUPPLEMENTAL.

Giggs, Rebecca. 2020. *Fathoms: The World in the Whale*. Melbourne: Scribe.

Goldfarb, Ben. 2019. *Eager: The Surprising, Secret Life of Beavers and Why They Matter*. White River Junction, VT: Chelsea Green.

Gravendeel, B., G. A. de Groot, M. Kik, K. Beentjes, H. Bergman, R. Caniglia, H. Cremers, et al. 2013. "The First Wolf Found in the Netherlands in 150 Years Was the Victim of a Wildlife Crime—WUR." *Lutra* 56 (2): 93–109. https://www.wur.nl/en/Publication-details.htm?publicationId=publication-way-343536323630.

Griscom, Bronson W., Justin Adams, Peter W. Ellis, Richard A. Houghton, Guy Lomax, Daniela A. Miteva, William H. Schlesinger, et al. 2017. "Natural Climate Solutions." *Proceedings of the National Academy of Sciences of the United States of America* 114 (44): 11645–11650. https://doi.org/10.1073/PNAS.1710465114/-/DCSUPPLEMENTAL.

Guardian, The. 2019. "Belgium's First Sighted Wolf in a Century Feared Killed by Hunters," October 2.

Guarino, Julia. 2013. "Tribal Advocacy and the Art of Dam Removal: The Lower Elwha Klallam and the Elwha Dams." *American Indian Law Journal* 2 (1): 114–145.

Gustafson, Richard G., Robin S. Waples, James M. Myers, Laurie A. Weitkamp, Gregory J. Bryant, Orlay W. Johnson, and Jeffrey J. Hard. 2007. "Pacific Salmon Extinctions: Quantifying Lost and Remaining Diversity." *Conservation Biology* 21 (4): 1009–1020. https://doi.org/10.1111/J.1523-1739.2007.00693.X.

Gutiérrez, R. J., M. Cody, S. Courtney, and Alan B. Franklin. 2007. "The Invasion of Barred Owls and Its Potential Effect on the Spotted Owl: A Conservation Conundrum." *Biological Invasions* 9 (2): 181–196. https://doi.org/10.1007/S10530-006-9025-5/TABLES/1.

Hagenbuch, Brian. 2021. "Off the Hook: Slinky Pots Revolutionize Alaska's Blackcod Fishery." *National Fisherman*, May.

Hedrick, Philip W. 2009. "Conservation Genetics and North American Bison (*Bison bison*)." *Journal of Heredity* 100 (4): 411–420. https://doi.org/10.1093/JHERED/ESP024.

Hershey, Katherine T., E. Charles Meslow, and Fred L. Ramsey. 1998. "Characteristics of Forests at Spotted Owl Nest Sites in the Pacific Northwest." *Journal of Wildlife Management* 62 (4): 1398. https://doi.org/10.2307/3802006.

Hood, W. Gregory. 2012. "Beaver in Tidal Marshes: Dam Effects on Low-Tide Channel Pools and Fish Use of Estuarine Habitat." *Wetlands* 32 (3): 401–410. https://doi.org/10.1007/S13157-012-0294-8.

Hughes, Terry P., James T. Kerry, Andrew H. Baird, Sean R. Connolly, Andreas Dietzel, C. Mark Eakin, Scott F. Heron, et al. 2018. "Global Warming Transforms Coral Reef Assemblages." *Nature* 556 (7702): 492–496. https://doi.org/10.1038/s41586-018-0041-2.

Huisman, N. 2018. "First Confirmed Wolf in Belgium in a Hundred Years." Wilderness Society. https://wilderness-society.org/first-confirmed-wolf-belgium-100-years.

Irvine, Amy, and Ruby McHarg. 2021. "Retriever of Souls." *Orion Magazine*, January 20.

Jakes, Andrew F., Paul F. Jones, L. Christine Paige, Renee G. Seidler, and Marcel P. Huijser. 2018. "A Fence Runs through It: A Call for Greater Attention to the Influence of Fences on Wildlife and Ecosystems." *Biological Conservation* 227 (June): 310–318. https://doi.org/10.1016/j.biocon.2018.09.026.

Jameson, Ronald, Karl Kenyon, Ancel M Johnson, and Howard M Wight. 1982. "History and Status of Translocated Sea Otter Populations in North America." *Wildlife Society Bulletin* 10 (2): 100–107. https://jstor.org/stable/3781726.

Jenkins, Julianna M. A., Damon B. Lesmeister, J. David Wiens, Jonathan T. Kane, Van R. Kane, and Jake Verschuyl. 2019. "Three-Dimensional Partitioning of Resources by Congeneric Forest Predators with Recent Sympatry." *Scientific Reports* 9 (1): 1–10. https://doi.org/10.1038/s41598-019-42426-0.

Joly, Kyle, Eliezer Gurarie, Mathew S. Sorum, Petra Kaczensky, Matthew D. Cameron, Andrew F. Jakes, Bridget L. Borg, et al. 2019. "Longest Terrestrial Migrations and Movements around the World." *Scientific Reports* 9 (1): 1–10. https://doi.org/10.1038/s41598-019-51884-5.

Keefer, Matthew L., Gregory A. Taylor, Douglas F. Garletts, Chad K. Helms, Greg A. Gauthier, Todd M. Pierce, and Christopher C. Caudill. 2012. "Reservoir Entrapment and Dam Passage Mortality of Juvenile Chinook Salmon in the Middle Fork Willamette River." *Ecology of Freshwater Fish* 21 (2): 222–234. https://doi.org/10.1111/J.1600-0633.2011.00540.X.

Kent Wildlife Trust. 2011. "The Blean: Canterbury and Swale's Ancient Woodlands." Kentish Stour Countryside Partnership.

Kimmerer, Robin. 2020. *Braiding Sweetgrass: Indigenous Wisdom, Scientific Knowledge and the Teachings of Plants*. New York: Penguin.

Kintisch, Eli. 2015. "'The Blob' Invades Pacific, Flummoxing Climate Experts." *Science* 348 (6230): 17–18. https://doi.org/10.1126/SCIENCE.348.6230.17/ASSET/82029CE6-09FD-4EE6-A20C-1FA12C5B442F/ASSETS/SCIENCE.348.6230.17.FP.PNG.

Klein, Christine A. 1999. "On Dams and Democracy." *Oregon Law Review* 78: 641.

Knize, Perri. 1993. "The Woman Who Runs with the Wolves." *Sports Illustrated*, October.

Knudsen, Curtis M., Steve L. Schroder, Craig A. Busack, Mark V. Johnston, Todd N. Pearsons, William J. Bosch, and David E. Fast. 2006. "Comparison of Life History Traits between First-Generation Hatchery and Wild Upper Yakima River Spring Chinook Salmon." *Transactions of the American Fisheries Society* 135 (4): 1130–1144. https://doi.org/10.1577/T05-121.1.

Krasińska, Małgorzata, and Zbigniew A. Krasiński. 2013. *European Bison: The Nature Monograph*. 2nd ed. Berlin: Springer.

Krause-Jensen, Dorte, and Carlos M. Duarte. 2016. "Substantial Role of Macroalgae in Marine Carbon Sequestration." *Nature Geoscience* 9 (10): 737–742. https://doi.org/10.1038/ngeo2790.

Kurlansky, Mark. 2020. *Salmon: A Fish, the Earth, and the History of Their Common Fate*. Ventura, CA: Patagonia Works.

Larson, Shawn E., James L. Bodkin, and Glenn R. VanBlaricom. 2015. *Sea Otter Conservation*. Amsterdam: Elsevier. https://doi.org/10.1016/C2013-0-18902-7.

Latham, A. David M., M. Cecilia Latham, Kyle H. Knopff, Mark Hebblewhite, and Stan Boutin. 2013. "Wolves, White-Tailed Deer, and Beaver: Implications of Seasonal Prey Switching for Woodland Caribou Declines." *Ecography* 36 (12): 1276–1290. https://doi.org/10.1111/j.1600-0587.2013.00035.x.

Lavery, Trish J., Ben Roudnew, Peter Gill, Justin Seymour, Laurent Seuront, Genevieve Johnson, James G. Mitchell, and Victor Smetacek. 2010. "Iron Defecation by Sperm Whales Stimulates Carbon Export in the Southern Ocean." *Proceedings of the Royal Society B: Biological Sciences* 277 (1699): 3527–3531. https://doi.org/10.1098/rspb.2010.0863.

Leslie, Jacques. 2019. "To Save Klamath River Salmon, Shut Down the Hatcheries." *Los Angeles Times*, June 13.

Levin, Phillip S., Richard W. Zabel, and John G. Williams. 2001. "The Road to Extinction Is Paved with Good Intentions: Negative Association of Fish Hatcheries with Threatened Salmon." *Proceedings of the Royal Society B: Biological Sciences* 268 (1472): 1153–1158. https://doi.org/10.1098/RSPB.2001.1634.

Liermann, Martin, George Pess, Mike McHenry, John McMillan, Mel Elofson, Todd Bennett, and Raymond Moses. 2017. "Relocation and Recolonization of Coho Salmon in Two Tributaries to the Elwha River: Implications for Management and Monitoring." *Transactions of the American Fisheries Society* 146 (5): 955–966. https://doi.org/10.1080/00028487.2017.1317664.

Livezey, Kent B. 2009a. "Range Expansion of Barred Owls, Part I: Chronology and Distribution." *American Midland Naturalist* 161 (1): 49–56. https://doi.org/10.1674/0003-0031-161.1.49.

Livezey, Kent B. 2009b. "Range Expansion of Barred Owls, Part II: Facilitating Ecological Changes." *American Midland Naturalist* 161 (2): 323–349.

Lockyer, Christina. 2007. "All Creatures Great and Smaller: A Study in Cetacean Life History Energetics." *Journal of the Marine Biological Association of the United Kingdom* 87 (4): 1035–1045. https://doi.org/10.1017/S0025315407054720.

Lopez, B. H. 1979. *Of Wolves and Men*. New York: Simon and Schuster.

Loy, A., P. Genov, M. Galfo, M. G. Jacobone, and A. Vigna Taglianti. 2008. "Cranial Morphometrics of the Apennine Brown Bear (*Ursus arctos marsicanus*) and Preliminary Notes on the Relationships with Other Southern European Populations." *Italian Journal of Zoology* 75 (1): 67–75. https://doi.org/10.1080/11250000701689857.

MacPhearson, J., C. Carter, S. Devillard, R. Kennerly, S. Rouette, and M. Hudson. 2019. "A Preliminary Feasibility Assessment for the Reintroduction of the European Wildcat to England and Wales." Vincent Wildlife Trust.

Mapes, Lynda, and Steve Ringman. 2013. *Elwha: A River Reborn*. Seattle, WA: The Seattle Times and The Mountaineer Books.

Marris, Emma. 2009. "Conservation: The Genome of the American West." *Nature* 457: 950–952.

Marris, Emma. 2015. "Handle with Care." *Orion Magazine*, April.

Marris, Emma. 2016. "Mysterious Origin of European Bison Revealed Using DNA and Cave Art." *Nature*.

Marris, Emma. 2021. *Wild Souls: Freedom and Flourishing in the Non-Human World*. New York: Bloomsbury.

Masson-Delmotte, V., et al. 2021. *Climate Change 2021: The Physical Science Basis. Contribution of Working Group I to the Sixth Assessment Report of the Intergovernmental Panel on Climate Change*. Cambridge: Cambridge University Press.

McIntyre, R. 2019. *The Rise of Wolf 8: Witnessing the Triumph of Yellowstone's Underdog*. Vancouver, BC: Greystone Books.

McIntyre, R. 2020. *The Reign of Wolf 21: The Saga of Yellowstone's Legendary Druid Pack*. Vancouver, BC: Greystone Books.

McIntyre, R. 2021. *The Redemption of Wolf 302: From Renegade to Yellowstone Alpha Male*. Vancouver, BC: Greystone Books.

McKinstry, Mark C., Paul Caffrey, and Stanley H. Anderson. 2001. "The Importance of Beaver to Wetland Habitats and Waterfowl in Wyoming." *JAWRA: Journal of the American Water Resources Association* 37 (6): 1571–1577. https://doi.org/10.1111/J.1752-1688.2001.TB03660.X.

McLeish, Todd. 2018. *Return of the Sea Otter: The Story of the Animal That Evaded Extinction on the Pacific Coast*. Seattle, WA: Sasquach Books.

McPhee, John. 2002. *The Founding Fish*. New York: Farrar, Straus and Giroux.

Meriggi, A., A. Brangi, L. Schenone, D. Signorelli, and P. Milanesi. 2011. "Changes of Wolf (*Canis lupus*) Diet in Italy in Relation to the Increase of Wild Ungulate Abundance." *Ethology Ecology and Evolution* 23 (3): 195–210. https://doi.org/10.1080/03949370.2011.577814.

Mills, E. A. 1913. *In Beaver World*. Boston: Houghton Mifflin.

Mills, L. Scott, Michael E. Soulé, and Daniel F. Doak. 1993. "The Keystone-Species Concept in Ecology and Conservation Management and Policy Must Explicitly Consider the Complexity of Interactions in Natural Systems." *BioScience* 43 (4): 219–224. https://doi.org/10.2307/1312122.

Mintz, D. 2021. "The Slinky Pot." Pacific Fishing. https://www.pacificfishing.com/featured_stories/0321_story2.html.

Monbiot, George. 2013. *Feral: Searching for Enchantment on the Frontiers of Rewilding*. London: Allen Lane.

Mooallem, Jon. 2013. *Wild Ones: A Sometimes Dismaying, Weirdly Reassuring Story about Looking at People Looking at Animals in America*. New York: Penguin Press.

Nash, R. 1967. *Wilderness and the American Mind*. New Haven, CT: Yale University Press.

Needham, Robert J., Martin Gaywood, Angus Tree, Nick Sotherton, Dylan Roberts, Colin W. Bean, and Paul S. Kemp. 2021. "The Response of a Brown Trout (*Salmo trutta*) Population to Reintroduced Eurasian Beaver (*Castor fiber*) Habitat Modification." *Canadian Journal of Fisheries and Aquatic Sciences* (October): 1–11. https://doi.org/10.1139/CJFAS-2021-0023/SUPPL_FILE/CJFAS-2021-0023SUPPLB.DOCX.

Neff, Bryan D., Shawn R. Garner, Ian A. Fleming, and Mart R. Gross. 2015. "Reproductive Success in Wild and Hatchery Male Coho Salmon." *Royal Society Open Science* 2 (8). https://doi.org/10.1098/RSOS.150161.

Nijhuis, Michelle. 2021. *Beloved Beasts: Fighting for Life in an Age of Extinction*. New York: Norton.

NL Times. 2021. "Thousands of Wild Boar and Deer to Be Shot This Year in Veluwe." July 9.

O'Brien, D. 2017. *Great Plains Bison*. Lincoln: University of Nebraska Press.

O'Sullivan, John. 1845. "Annexation." *The United States Magazine and Democratic Review* 17 (85): 5–10. https://pdcrodas.webs.ull.es/anglo/OSullivanAnnexation.pdf.

Peninsula Daily News. 2019. "Large Coho Salmon Smolts Found Exiting Lake Sutherland." September 6.

Penteriani, Vincenzo, and Mario Melleti, eds. 2020. *Bears of the World: Ecology, Conservation and Management*. Cambridge: Cambridge University Press. https://doi.org/10.1017/9781108692571.

Pershing, Andrew J., Line B. Christensen, Nicholas R. Record, Graham D. Sherwood, and Peter B. Stetson. 2010. "The Impact of Whaling on the Ocean Carbon Cycle: Why Bigger Was Better." *PLoS ONE* 5 (8): 1–10. https://doi.org/10.1371/journal.pone.0012444.

Pirastru, Monica, Paolo Mereu, Laura Manca, Daniela Bebbere, Salvatore Naitana, and Giovanni G. Leoni. 2021. "Anthropogenic Drivers Leading to Population Decline and Genetic Preservation of the Eurasian Griffon Vulture (*Gyps fulvus*)." *Life* 11 (10): 1038. https://doi.org/10.3390/LIFE11101038.

Pooley, Simon, Saloni Bhatia, and Anirudhkumar Vasava. 2021. "Rethinking the Study of Human–Wildlife Coexistence." *Conservation Biology* 35 (3): 784–793. https://doi.org/10.1111/COBI.13653.

Post, A. 2008. "Why Fish Need Trees and Trees Need Fish." Alaska Fish & Wildlife News, Alaska Department of Fish and Game, November. https://www.adfg.alaska.gov/index.cfm?adfg=wildlifenews.view_article&articles_id=407.

Ražen, Nina, Alessandro Brugnoli, Chiara Castagna, Claudio Groff, Petra Kaczensky, Franci Kljun, Felix Knauer, et al. 2016. "Long-Distance Dispersal Connects Dinaric-Balkan and Alpine Grey Wolf (*Canis lupus*) Populations." *European Journal of Wildlife Research* 62 (1): 137–142. https://doi.org/10.1007/s10344-015-0971-z.

Reinhardt, Ilka, Gesa Kluth, Carsten Nowak, Claudia A. Szentiks, Oliver Krone, Hermann Ansorge, and Thomas Mueller. 2019. "Military Training Areas Facilitate the Recolonization of Wolves in Germany." *Conservation Letters* 12 (3): e12635. https://doi.org/10.1111/CONL.12635.

Roman, Joe, James A. Estes, Lyne Morissette, Craig Smith, Daniel Costa, James McCarthy, J. B. Nation, Stephen Nicol, Andrew Pershing, and Victor Smetacek. 2014. "Whales as Marine Ecosystem Engineers." *Frontiers in Ecology and the Environment* 12 (7): 377–385. https://doi.org/10.1890/130220.

Roman, Joe, and James J. McCarthy. 2010. "The Whale Pump: Marine Mammals Enhance Primary Productivity in a Coastal Basin." *PLoS ONE* 5 (10). https://doi.org/10.1371/journal.pone.0013255.

Rose, K., and E. Kwong. 2018. "Sitka Tribe Sues State, Claiming Mismanagement of Herring Fishery." KCAW.org, December 18.

Russell, Osborne. 1965. *Journal of a Trapper*. Edited by Aubrey L Haynes. Lincoln: University of Nebraska Press.

Sadin, Paul, Dawn Vogel, and Heather Lee Miller. 2011. "An Interpretive History of the Elwha River Valley and the Legacy of Hydropower on Washington's Olympic Peninsula." Historical Research Associates, Seattle, WA, for the National Park Service, Port Angeles, WA.

Safina, Carl. 2015. *Beyond Words: What Animals Think and Feel*. New York: Henry Holt.

Salomon, Anne K., Jennifer M. Burt, K̲ii'iljuus Barb J. Wilson, Nicky Roberts, Laurie Wood, and Leah Honka. 2018. "Visioning the Future of Kelp Forest, Sea Otter and Human Interactions." Workshop summary report, School of Resource and Environmental Management, Simon Fraser University, Burnaby, British Columbia, Canada.

Salviamo l'Orso (SLO). 2020. "Bear-Smart Community Best Practices Manual." Salviamo l'Orso, Montesilvano, Italy.

Sauer, John R., Kenneth V. Rosenberg, Adriaan M. Dokter, Peter J. Blancher, Adam C. Smith, Paul A. Smith, J. C. Stanton, et al. 2019. "Decline of the North American Avifauna." *Science* 366 (6461): 120–124. https://doi.org/10.1126/SCIENCE.AAW1313.

Savoca, Matthew S., Max F. Czapanskiy, Shirel R. Kahane-Rapport, William T. Gough, James A. Fahlbusch, K. C. Bierlich, Paolo S. Segre, et al. 2021. "Baleen Whale Prey

Consumption Based on High-Resolution Foraging Measurements." *Nature* 599 (7883): 85–90. https://doi.org/10.1038/S41586-021-03991-5.

Schiel, David R., and Michael S. Foster. 2015. *The Biology and Ecology of Giant Kelp Forests.* Berkeley: University of California Press. https://doi.org/10.1525/9780520961098.

Schmitz, Oswald. 2016. "How 'Natural Geoengineering' Can Help Slow Global Warming." Yale Environment 360, Yale School of the Environment. https://e360.yale.edu/features/how_natural_geo-engineering_can_help_slow_global_warming.

Schuette, Paul, Scott Creel, and David Christianson. 2013. "Coexistence of African Lions, Livestock, and People in a Landscape with Variable Human Land Use and Seasonal Movements." *Biological Conservation* 157 (January): 148–154. https://doi.org/10.1016/J.BIOCON.2012.09.011.

Scott, J. M., D. D. Goble, J. A. Wiens, D. S. Wilcove, M. Bean, and T. Male. 2005. "Recovery of Imperiled Species under the Endangered Species Act: The Need for a New Approach." *Frontiers in Ecology and the Environment* 3 (7): 383–389. https://doi.org/10.1890/1540-9295(2005)003[0383:ROISUT]2.0.CO;2.

Senn, Helen V., Muhammad Ghazali, Jennifer Kaden, David Barclay, Ben Harrower, Ruairidh D. Campbell, David W. Macdonald, and Andrew C. Kitchener. 2019. "Distinguishing the Victim from the Threat: SNP-Based Methods Reveal the Extent of Introgressive Hybridization between Wildcats and Domestic Cats in Scotland and Inform Future *in situ* and *ex situ* Management Options for Species Restoration." *Evolutionary Applications* 12 (3): 399–414. https://doi.org/10.1111/EVA.12720.

Seton, Ernest Thomas. 1929. *Lives of Game Animals.* New York: Doubleday, Duran.

Shaw, James H. 1995. "How Many Bison Originally Populated Western Rangelands?" *Rangelands* 17 (5): 148–150.

Shoemaker, N. 2019. "Oil, Spermaceti, Ambergris, and Teeth: Products of the Nineteenth Century Pacific Sperm Whale Industry." *RCC Perspectives* 5: 17–22.

Silverstein, Alvin, Virginia Silverstein, and Robert Silverstein. 1995. *The Sea Otter.* Brookfield, CT: Millbrook Press.

Smith, Craig R., Adrian G. Glover, Tina Treude, Nicholas D. Higgs, and Diva J. Amon. 2015. "Whale-Fall Ecosystems: Recent Insights into Ecology, Paleoecology, and Evolution." *Annual Review of Marine Science* 7 (January): 571–596. https://doi.org/10.1146/ANNUREV-MARINE-010213-135144.

Smith, Douglas W., Daniel R. Stahler, and Daniel R. MacNulty, eds. 2020. *Yellowstone Wolves: Science and Discovery in the World's First National Park*, 13–25. Chicago: University of Chicago Press.

Smith, J. 1910. *Travels and Works of Captain John Smith.* Edited by E. Arber and A. G. Bradley. Edinburgh, Scotland: John Grant.

Smith, Joshua G., Joseph Tomoleoni, Michelle Staedler, Sophia Lyon, Jessica Fujii, and M. Tim Tinker. 2021. "Behavioral Responses across a Mosaic of Ecosystem States

Restructure a Sea Otter–Urchin Trophic Cascade." *Proceedings of the National Academy of Sciences of the United States of America* 118 (11): 1–7. https://doi.org/10.1073/pnas.2012493118.

Soubrier, J., G. Gower, K. Chen, S. Richards, and A. Cooper. 2016. "Early Cave Art and Ancient DNA Reveal the Origin of European Bison." *Nature Communications* 7: 107.

Soulé, Michael E. 1985. "What Is Conservation Biology?" *BioScience* 35 (11): 727–734. https://doi.org/10.2307/1310054.

Spokesman-Review. 2009. "Wolves Kill 120 Sheep Near Dillon Montana," August 28, 2009.

Steffen, Will, P. J. Crutzen, and J. R. McNeill. 2007. "The Anthropocene: Are Humans Now Overwhelming the Great Forces of Nature." *Ambio-Journal of Human Environment Research and Management* 36 (8): 614–621.

Stroupe, Sam, David Forgacs, Andrew Harris, James N. Derr, and Brian W. Davis. 2022. "Genomic Evaluation of Hybridization in Historic and Modern North American Bison (*Bison bison*)." *Scientific Reports* 12 (1): 1–11.

Swann, Kirsten. 2017. "Eklutna River Restoration Efforts Moving Forward." *Chugiak-Eagle River Star*, September 5.

Thoreau, H. D. 1980. *A Week on the Concord and Merrimack Rivers*. Princeton, NJ: Princeton University Press.

Thornton, Thomas F., and Harvey Kitka. 2015. "An Indigenous Model of a Contested Pacific Herring Fishery in Sitka, Alaska." *International Journal of Applied Geospatial Research* 6 (1): 94–117. https://doi.org/10.4018/ijagr.2015010106.

Tonra, Christopher M., Kimberly Sager-Fradkin, Sarah A. Morley, Jeffrey J. Duda, and Peter P. Marra. 2015. "The Rapid Return of Marine-Derived Nutrients to a Freshwater Food Web Following Dam Removal." *Biological Conservation* 192 (December): 130–134. https://doi.org/10.1016/J.BIOCON.2015.09.009.

Turley, C. 2002. "The Importance of 'Marine Snow.'" *Microbiology Today* 29: 177–179.

U.S. Department of Agriculture, Animal and Plant Health Inspection Service (USDA-APHIS). 2020. "USDA Animal and Plant Health Inspection Service (Wildlife Services Division)." USDA-APHIS, Washington, DC.

van Klink, Roel, Jasper L. Ruifrok, and Christian Smit. 2016. "Rewilding with Large Herbivores: Direct Effects and Edge Effects of Grazing Refuges on Plant and Invertebrate Communities." *Agriculture, Ecosystems and Environment* 234: 81–97. https://doi.org/10.1016/j.agee.2016.01.050.

Walters, J. R., S. R. Derrickson, D. M. Fry, S. M. Haig, J. M. Marzluff, and J. M. Wunderle Jr. 2008. "Status of the California Condor and Efforts to Achieve Its Recovery." American Ornithologists' Union and Audubon California.

Wassman, P., M. Biuw, and T. Haug. 2021. "A Critical Evaluation of Whales as Ecosystem Engineers." International Whaling Commission. Cambridge, UK.

Welch, C. 2009. "The Spotted Owl's New Nemesis." *Smithsonian Magazine*, January.

Whipple, Alexa. 2019. "Riparian Resilience in the Face of Interacting Disturbances: Understanding Complex Interactions between Wildfire, Erosion, and Beaver (*Castor canadensis*) in Grazed Dryland Riparian Systems of Low Order Streams in North Central Washington State, USA." Master's thesis, EWU Masters Thesis Collection, Eastern Washington University.

Wiens, J. David, Katie M. Dugger, J. Mark Higley, Damon B. Lesmeister, Alan B. Franklin, Keith A. Hamm, Gary C. White, et al. 2021. "Invader Removal Triggers Competitive Release in a Threatened Avian Predator." *Proceedings of the National Academy of Sciences of the United States of America* 118 (31). https://doi.org/10.1073/pnas.2102859118.

Wiens, J. D., K. M. Dugger, D. B. Lesmeister, K. E. Dilione, and D. C. Simon. 2020. "Effects of Barred Owl (*Strix varia*) Removal on Population Demography of Northern Spotted Owls (*Strix occidentalis caurina*) in Washington and Oregon." 2019 Annual Report Prepared in Cooperation with the U.S. Fish and Wildlife Service, Bureau of Land Management. https://doi.org/https://doi.org/10.3133/ofr20201089.

Williams, P. 2021. "Furs and Futures." *First Alaskans.*

Wilmers, Christopher C., James A. Estes, Matthew Edwards, Kristin L. Laidre, and Brenda Konar. 2012. "Do Trophic Cascades Affect the Storage and Flux of Atmospheric Carbon? An Analysis of Sea Otters and Kelp Forests." *Frontiers in Ecology and the Environment* 10 (8): 409–415. https://doi.org/10.1890/110176.

Wilson, Grant S. 2013. "Murky Waters: Ambiguous International Law for Ocean Fertilization and Other Geoengineering." *SSRN Electronic Journal*, August. https://doi.org/10.2139/SSRN.2312755.

Wohl, Ellen. 2013. "Landscape-Scale Carbon Storage Associated with Beaver Dams." *Geophysical Research Letters* 40 (14): 3631–3636. https://doi.org/10.1002/GRL.50710.

Wood, Judith Hebbring. 2000. "The Origin of Public Bison Herds in the United States." *Wicazo Sa Review* 15 (1): 157–182.

Woodford, Riley. 2003. "Sperm Whales Awe and Vex Alaska Fishermen." Alaska Fish and Wildlife News, August. http://www.adfg.alaska.gov/index.cfm?adfg=wildlifenews.view_article&articles_id=61.

Würsig, Bernd G., J. G. M. Thewissen, and Kit M. Kovacs. 2018. *Encyclopedia of Marine Mammals.* 3rd ed. Aalborg, Denmark: Elsevier.

Xiu, Peng, Andrew C. Thomas, and Fei Chai. 2014. "Satellite Bio-Optical and Altimeter Comparisons of Phytoplankton Blooms Induced by Natural and Artificial Iron Addition in the Gulf of Alaska." *Remote Sensing of Environment* 145 (April): 38–46. https://doi.org/10.1016/J.RSE.2014.02.004.

INDEX